Friday March 8th
Evening 9 o'clock

...several articles to go by
and mother made some seeds
they did not go, and we
as if you would have to
wrote thinking no one was

The Aurora's boats whaling
a school of sperm whales
1858
George E Mills

...Whaling Song

Fayal Sept 11th /51

Dear Brother

I have but a moment to spare so you must be satisfied with a few lines, Elizabeths letter has been received by the Jo, which was very pleas't to us indeed, and Malvina will answer it as soon as she can sit long enough to write, I have sent a few Oranges to you by the Bark Savannah Captn Lunt, and Seth, will forward them if nothing happ'n, should they arrive in good order please send Mother Marshall a half dozen, (Malvina is very well except one sore breast, which she is in hopes will soon get well) the Babe is very well and we are in hopes it will continue so, Malvina is very anxious to get home with her as she has been to the Worlds Fair after her instead of sending to Boston for her) on the 8th there was a Dutch Brig arrived here that was dismasted in the Hurricane that we had, they had a dreadfull time, she got thrown down so that the water went

George E Mills

Nellie's Boat
1858
Bark Aurora On the
For George E Mills

Nellie... Dearest Nellie

We are pleased to inscribe this book

Marsha Hall Brown

Author

Tom Hall

Contributing Editor

Dawn Secord

Illustrator of Pastels and Drawings

Especially for Hunter and Logan
on this special day on main street
Nantucket

Marsha
July 27, 2019

Under Sun, Stars and Sails

A Whaling Family's Life at Sea

Helen Marshall ambrotype by William G. Helsby, Valparaiso, Chile, 1860. *Helsby is noted for his images of the Hoa Hakananai'a statue on Easter Island and the first existing photographs of Tahiti. Helsby exhibits are in the British Museum and the Getty Museum.* – Nantucket Historical Association

Under Sun, Stars and Sails

A Whaling Family's Life at Sea

Marsha Hall Brown

with

Tom Hall

Spinner Publications, Inc.

New Bedford, Massachusetts

FRONT COVER: **"New Bedford Arrival, 1861."** Before sailing home to Nantucket, Malvina Marshall waves to her welcoming family as the *Aurora* arrives at New Bedford. Malvina's silk taffeta dress is in the collection of the Nantucket Historical Association. – Pastel by Dawn Secord

BACK COVER: **The *Sea Queen* of Westport, Massachusetts.** – New Bedford Whaling Museum
Captain Joseph Marshall. – Nantucket Historical Association

FRONT AND BACK ENDPAPERS: Letters, envelopes, and artifacts in the Marshall, Pinkham, and Crosby family collections at the Nantucket Historical Association; logbook pages from the New Bedford Whaling Museum.

Spinner Publications, Inc., New Bedford, Massachusetts 02740

Printed in the United States of America.

Library of Congress Cataloging-in-Publication Data

Names: Brown, Marsha Hall, author.
Title: Under Sun, Stars and Sails : A Whaling family's life at sea / by Marsha Hall Brown.
Description: New Bedford, Massachusetts : Spinner Publications, Inc., 2019. |
Includes bibliographical references and index.
Identifiers: LCCN 2019010433 | ISBN 0932027334 (cloth)
Subjects: LCSH: Marshall, Joseph, 1811-1879. | Marshall, Malvina Fitzalan Pinkham, 1820-1885. |
Marshall, Helen, 1851-1939. | Aurora (Bark) | Voyages and travels. | Seafaring life--History--19th century. |
Whaling--History--19th century. | Nantucket (Mass.)--Biography.
Classification: LCC G226.M32 B76 2019 | DDC 910.4/509255 [B] --dc23
LC record available at https://lccn.loc.gov/2019010433

Dedicated to Florence Farrier Hall, 1900–1990

For keeping the stories about Helen Marshall and her family alive for today's readers and for saving and protecting the numerous Marshall family artifacts that are now archived in the collection of the Nantucket Historical Association and the Hall-Brown Private Collections.

Spinner Publications and Marsha Hall Brown gratefully acknowledge the Nantucket Historical Association as the primary underwriter for *Under Sun, Stars and Sails.*

We also thank these individual sponsors for their generous gifts:

Captain's Circle Sponsors

Howelston History Fund, research grant

In Memoriam: Rev. Dr. Robert E. Farrier and
Helen Locke Farrier, Squam Head, Nantucket

Quarter Deck Sponsors

Brown Family Trust
In Honor of the Builders of the *Sea Queen* and *Aurora*:
Alonzo Matthews and John Mashow

Neto Insurance Company, Manuel Neto

Shipkeeper Sponsors

Lucy Alexandria Bly
Leo P. Dalbec
Decendants of Whaling Masters, Inc.

People will not look forward to posterity who never look backward to their ancestors. –Edmund Burke

Contents

List of Illustrations

Foreword

Marsha Hall Brown brings us on an exhilarating and vicarious voyage around the world as she pieces together through meticulous research the unusual childhood of Helen Marshall. In doing so, we are treated to an intimate look at a seafaring family and equally engrossed in the global business that was Yankee whaling. From Nantucket, bluff bowed ships embarked on harrowing voyages that often lasted years. A relentless pursuit of the Leviathan brought these intrepid islanders to far flung shores. So it was with the Marshall family as we follow their trek across the Atlantic and Pacific Oceans and around the Horn and back twice before Helen was 10 years old. In doing so, we are rewarded with an accounting of what life was like aboard a nineteenth-century whaler replete with the trials and tribulations, the toil and exertions, and the joys and exhilarations of life at sea. Brown takes us to the most exotic places on Earth as the Marshall family meets consuls, diplomats, envoys, plus a prince, king, and queen en route.

Nantucket was a maritime powerhouse in the latter half of the eighteenth and first half of the nineteenth centuries, famed for fostering the most accomplished whaling captains in the world, for sending ships to the far reaches of the Pacific, and for bringing great wealth to the small island in the North Atlantic. A few intrepid Nantucket women began joining their sea-captain husbands on multi-year whaling voyages, leaving the comforts of home and extended family for the adventures of a lifetime. Their husbands benefited from continual family time, while crews often doted on any children. Malvina Marshall sailed on the *Sea Queen* of Westport, Massachusetts with her husband, Captain Joseph Marshall, on October 15, 1851. Six weeks later, their daughter, Helen was born on the island of Fayal, in the Azores. Helen spent formative years aboard whalers with her parents. Malvina brought a touch of domesticity to shipboard life, educating her daughter, and using her mathematical skills to help with navigation.

The Marshalls returned to Nantucket in 1861 where Helen continued her education on the island and at Vassar College. She set off on a Grand Tour of Europe in 1876 with her friend Ann Mitchell Macy, sister of famed astronomer Maria Mitchell. Upon her return, she taught at Nantucket High School and later at Norwich Free Academy in Connecticut, thus following a long and distinguished line of Nantucket women who excelled in educating this island's youth.

In 2017, Michael R. Harrison, the NHA's Obed Macy Director of Research and Collections, curated an excellent exhibition titled *"Dear Absent Ones": The Seafaring Pinkhams*. This exhibition looked at Nantucket seafaring through the eyes of the Pinkham and Marshall families. The exhibition would not have been possible were it not for the remarkable generosity of Marsha Hall Brown, as she had given the NHA numerous artifacts and nearly 200 letters relating to her Nantucket ancestors. The core of her donations relate to Joseph, Malvina, and Helen Marshall, some of which serve as illustrations in this book and reflect their time spent together on whaling voyages in the 1850s. Through family letters, correspondences, heirlooms, and artifacts, we are voyeurs into a period of time when Nantucket was undergoing a transformation. This tightly-written and fast-paced work will serve as an enduring tribute to one of Nantucket's notable families.

–James Russell, Gosnell Executive Director
Nantucket Historical Association

Introduction

The eighteenth-century age of sail in New England gave rise to an industry that would soon employ many thousands of Americans, create significant wealth among its ship owners and workers, and provide a myriad of products that would become essential to life in North America and the world.

According to traditional accounts, it began when Wampanoag Indians on Nantucket Island showed their white English neighbors how to use a small boat to chase and harpoon a small scrag whale that had ventured into one of the island's harbors. When the blubber was stripped off and boiled or "tried out" in large vats on the beach, the fat was reduced to a very useful oil. Further aided by instructions from the "whale citchers" on Cape Cod, the neophyte Nantucket whalemen introduced a number of innovations that enabled them to set out on whale hunts in the open ocean. They designed and built three-masted whaling ships that were large enough to carry brick furnaces and try pots on the decks and had room in the hold to store barrels of oil. Portable whaleboats—outfitted for the chase with oars, sails, rope, buckets, lanterns, and assorted tools and implements—were hung on davits above the deck and lowered into the water when whales were sighted. With these ships, boats, and try works, whalers became self-sufficient manufacturers at sea.

These skilled mariners soon turned their attention to studying the migratory habits of the right and sperm whales, and they practiced and improved their trade as they sailed to the far reaches of both the Atlantic and Pacific Oceans. Success was measured in barrels of oil; when their holds were full, the captains headed for home. Soon ships from Nantucket, New Bedford, and other New England ports began to arrange for delivery of these barrels directly to London, from where they would be shipped across the continent and used to light the cities of Europe. And it was not long before the streets of American cities were illuminated, seaports had lighthouses, and homes could enjoy not only whale oil lamps, but also the finest spermaceti candles.

Whale oil also became an essential lubricant for industrial manufacturing. Whale bones, baleen, and teeth were turned into kitchen implements, household tools, buttons, corset stays, umbrellas, and beautifully carved and etched scrimshaw art. Most prized of all parts of the whale was a secretion called ambergris, found in the abdomens of sperm whales, which provided the base for fine perfume and became as valuable as gold.

In the nineteenth century, after the long and difficult recovery from the American Revolution and the War of 1812 that had decimated the Nantucket fleet, the Nantucket Whale Fishery began to flourish once more. In 1825, seven hundred Nantucket ships were either hunting whales in the Pacific or being refitted at home. Ship owners and investors began demanding larger ships that could sustain longer voyages and carry larger cargoes. Such ships would commonly make voyages of three or four years or longer and had enough room for the captains to take their wives, and sometimes their children, to sea. And these were the developments that enabled the Marshall family to undertake the long voyages that provide the setting for this book.

Readers of all ages have long been fascinated by both fiction and nonfiction accounts of seagoing New Englanders—whether directly or only obliquely related to whaling. In both film and book form, *Moby-Dick, Mutiny on the Bounty, Two Years Before the Mast, The Loss of the Whaleship Essex*, and *Whale Hunt* are now succeeded by *In the Heart of the Sea*. Writers and directors skillfully employ

"South Sea Whale Fishery." *In this aquatint after an 1835 painting by Ambroise L. Garneray (1783–1857), the harpooner prepares to strike a lethal blow into a dying right whale. In the background, while another whaleboat tows a dead whale alongside the ship, men on board cut-in a whale and try out blubber into oil.* –New Bedford Whaling Museum

plots, settings, and characters relevant to whaling scenarios to create extraordinary levels of high adventure at sea. In addition, of course, expository texts examine every facet of whaling culture—navigation, geography, war, linguistic and cultural differences, loneliness, illness, accidents, ports of call, trade, acquisitions of food and water, and, naturally, the killing and the rendering of the whale.

The goal of this book, however, is not to focus on the drama, danger, and technical details of whaling, but to tell the story of the Marshall family of Nantucket during their voyages between 1851 to 1861. Excerpts from the unpublished writings of Malvina and Helen Marshall allow the reader to enter Helen's childhood world, while illustrations depict what that world looked like. The content of numerous letters and publications by and about Joseph and Malvina Marshall tell the story of how a ship's captain and his wife became true partners in the business of whaling while devoting themselves to each other and their unique family life with their young daughter. Subsumed under this primary theme are glimpses of the influence of Quaker beliefs and social mores that gave rise to gender equality in personal roles and a division of labor uncommon in the rest of America.

In essence, this book is intended for readers who love the sea and who are curious about the relationships that existed among those who made their living from it. While it may also serve as a source of supplemental reading on world history and geography, American studies, women's studies, and navigation, the most valuable aspect of this book is its unique and vivid depiction of family life aboard a whaling ship. Through the story of the Marshall family, *Under Sun, Stars and Sails* shines a light on the courageous women who braved the dangers and difficulties of life at sea in order to keep their families together. It portrays the captains who managed to maintain close, warm, and tender relationships with their loved ones while performing the cold, hard work of commanding a whaleship; and it depicts the daily life of parents and child whose primary reality is circumnavigating the globe amid surroundings that often belied the sophisticated and social lifestyle they had left behind in Nantucket.

Captain Joseph Marshall (1811–1879), ambrotype circa 1856. – Nantucket Historical Association

Chapter One

October 1851: Setting a Course

Malvina's husband, Joseph Marshall, was a master mariner wholly focused on his responsibility for the success of the *Sea Queen's* voyage to the whaling grounds beyond the treacherous waters of Cape Horn. With Joseph occupied with his duties, Malvina stood alone on the deck of the ship looking down at the unnaturally calm waters of the Atlantic, pondering how her life had changed since deciding to join Joseph at sea.

For four of the five years of her marriage, she had stayed at home on Nantucket while Joseph sailed to the Pacific in that most dangerous of occupations—whaling.

Certainly, she had known long separations before, since she was the daughter of Seth Pinkham, a successful whaling captain; she'd spent most of her youth waiting for her father's return from long voyages.

But since neither she nor her mother had ever gone to sea, Malvina knew little about the harsh conditions facing a woman on board ship. And she was not prepared for Joseph's change in demeanor as he assumed a firm command of the officers and crew. She'd never seen him so serious and brusque. At home, he'd always been adept at small talk, never at a loss for amusing anecdotes to tell at the many social gatherings they attended. And in private, he'd always spoken to her with touching sincerity, often telling her that she was the prettiest of the six Pinkham sisters.

But on this day she did not feel pretty; her soft, brown hair hung limp, free of its usual fashionable styling; she wore a plain, high-waisted cotton print mantua, made with her own skilled hands to provide for her expanding figure. Their first child was not due for more than a month, and by that time, the *Sea Queen* would be safely past the unpredictable weather and currents of the Atlantic.[1]

The voyage had begun on the fifteenth of October with great optimism and even celebration. The *Sea Queen* was a new ship, built by the owners specifically for Captain Marshall.[2] The sails, still sparkling white, were of the best canvas. The officers, though young, were well experienced, having sailed together before. And although it was the first whaling expedition for many of them, the crew members came from the eastern port cities of Massachusetts, spoke English, and were fit to meet the physical challenges of learning the ropes of a whaler.

The send-off had been memorable, with members of both the Marshall and Pinkham families traveling from Nantucket to New Bedford for the event.[3] Malvina's sister Elizabeth had even gone out on the pilot boat to watch for the last wave of Joseph's handkerchief before the *Sea Queen's* sails fell below the horizon.

The ship would follow the northern edge of the powerful Gulf Stream northeast toward Europe. This "highway in the sea" was the fastest route to the Azores. But being becalmed had already put them considerably behind schedule, creating unease among the officers and spreading superstitions among the crew. Was the present calm a harbinger of perils to come?

Regardless, with no regular shipboard duties required yet, there was ample time for the crew to be

Bark *Sea Queen* under sail. *On her maiden voyage (October 1851–April 1855), the* Sea Queen, *rigged as a brig under Captain Joseph Marshall, took 412 sperm whales, rendering 1,082 barrels of oil.* –From a painting by an unknown artist, New Bedford Whaling Museum

drilled in the many jobs they would soon enough be called upon to perform. Malvina could hear the officers bellowing orders to green-hand sailors as they practiced over and over the routine of launching and rowing the whale boats that were lowered from their davits and cast off from the ship.

Then, quite unexpectedly, Mr. Coffin, the second mate, appeared from behind the mizzenmast and, with quickened step and serious expression, approached Malvina, who was still deep in thought. He delivered his message to her with an unusually official voice. "Marm, the captain requires you to retire immediately to your quarters." He reached out with utmost courtesy, took her arm, and escorted her down the steep spiral of the companionway to the main stateroom. Although nothing seemed to be amiss, Malvina did not question the orders or the directions.

Matthews, Mashow & Company Shipyard,[5] 1850s. *Both the* Sea Queen *and the* Aurora *were built at this Padanaram Village shipyard in Dartmouth, Massachusetts. The etching provides a rare look into shipyards of the period. Alonzo Matthews had trained as a carpenter and John Mashow, a black man, had been born into slavery in South Carolina. Together they built more than 14 whaling barks and 19 schooners from 1846 to 1858.* – Etching from the 1858 map of the town of Dartmouth; information from *Sketches of Old Dartmouth, No. 2*

When she reached the quarters, she found the steward packing breakables and lashing the furniture and the trunks to the gunwale hooks. He explained that all hands were preparing for an impending storm.[4] Through the skylight, Malvina could see that darkness had suddenly surrounded the ship, even though it was still early in the day. For the first time in many days, she could hear the wind, and now it whipped through the luffing sails, whose beating canvas sounded like giant drums, and she heard the ship planks creaking loudly under the impact of the buffeting waves. Above, a change in tack had filled the lower sails with sudden force, heeling the *Sea Queen* sharply to starboard, shifting cargo above and below, and scattering the furnishings within.

Malvina inched her way along the slanting bulkhead until she could steady herself on the posts of the bed. Joseph had prepared her for emergencies such as this, instructing her to climb into the gimballed bed and brace herself against the forward posts. Once there, she told the steward that she was quite safe and gave him leave to join the crew topside.

The sound of quickened boot steps on the deck above told her that the men were rushing to get the sails down. Knowing that Joseph would remain on the quarterdeck as long as the ship was in danger from the storm, Malvina hurried to light the cabin's main hanging lantern at the first sign that the *Sea Queen* had become more stable, then poured herself some water. Any thought of food was out of the question.

Tropical squalls could be unpredictable as well as violent, and minutes later, the ship felt the full force of the wind and heavy-driving rain. The ship's rolling motion brought on a bout of sea-sickness for Malvina. Then came a violent fore-and-aft pitching as the ship mounted the steep crests of successive waves and then crashed into the troughs between them before rising again. Malvina hung on to the bed rails and watched the ominous flickering shadows cast by the wildly swinging lantern.

Many hours passed before the steward arrived with tea and bread, and it was hours more before Joseph appeared. He was pale, slump-shouldered, and his clothes were soaked through. He didn't share his concerns with his wife, but his experience with the Atlantic's volatility, along with the harsh and unstable recent winds and dismal prognosis of the ship's barometer, caused him to fear that the rest of the trip was not going to be routine. His quiet presence reassured Malvina nonetheless. She braved the perpetual motion of the cabin to tend to his comforts, then said a silent prayer of thanks that they were together.

But, as it would turn out, Joseph's concerns were well founded. While the weather was clear enough to make sightings of the island of Flores at daybreak on November 6, by midday, a storm hit with unusual fury. A whaleboat was knocked off its crane, improperly secured crates and supplies on the deck were swept overboard, and the green-hands became seasick and frightened. The captain shouted orders in a voice the men had not heard before; it was as threatening as the gale force winds and beating torrents of rain. All hands were needed to take in all canvas except the mainsail and the storm staysail. And it had to be done fast.

Three days later, a heavy gale set in and would not let up, badly frightening first mate William Snow,[6] who was at the helm. With seas breaking over the deck, he ordered every rag except the storm staysail to be brought down, all deck equipment to be re-lashed, and the waist boat to be bound to the try works. He and his watch crew got their axes ready, expecting to see the masts go overboard at any moment. A second huge wave hit the ship and raced across the deck as high as the top of the hurricane house, completely covering the mate and stoving in the skylight.

The captain, who stood beside the mainmast watching for any signs of danger to the yards, then raced to the companionway and down to the main stateroom. He found Malvina drenched from the water pouring into the cabin from above. Up to his knees in it, Joseph bored holes through the deck to let the water flow out.

He'd spent twenty-three years at sea and had never seen the Atlantic or Pacific so violent; he'd "never seen the sea run so high or heard the wind blow more than half as hard."[7]

The southeast wind blew for thirteen days with successive gales and one hurricane, until finally, on November 22, the sky cleared and a fine breeze allowed the crew to set every sail that would draw. By afternoon, the unique yet familiar outline of the Azorean Islands of St. George, Pico, and Fayal were in sight. It had taken eighteen days from the sighting of Flores to the entrance of the Bay of Horta on Fayal.

In comparison to the loss and devastation suffered by numerous ships from many countries across a huge area of the Atlantic, the *Sea Queen* suffered minimal damage. Only two of the thirty-man crew were too ill to work. While some of the crates on the forecastle deck and five of the animals housed between the cookhouse and the mainmast were swept overboard, the food supply was intact, for the most part, and the cook was back to work providing regular meals. Miraculously, none of the sails were damaged.

The *Sea Queen* had survived as the result of a skillful and experienced captain, the bravery and resolve of the crew, a tight and well-built new ship, and what Joseph declared to be "the grace of the almighty."

Port of Horta on the island of Fayal (left) and the volcanic island of Pico, Azores, circa 1848.[8] *Outward bound New England whaleships frequently made the Azores their first port of call for provisions and additional crew.* – Excerpt from Russell & Purrington's panorama,[9] New Bedford Whaling Museum

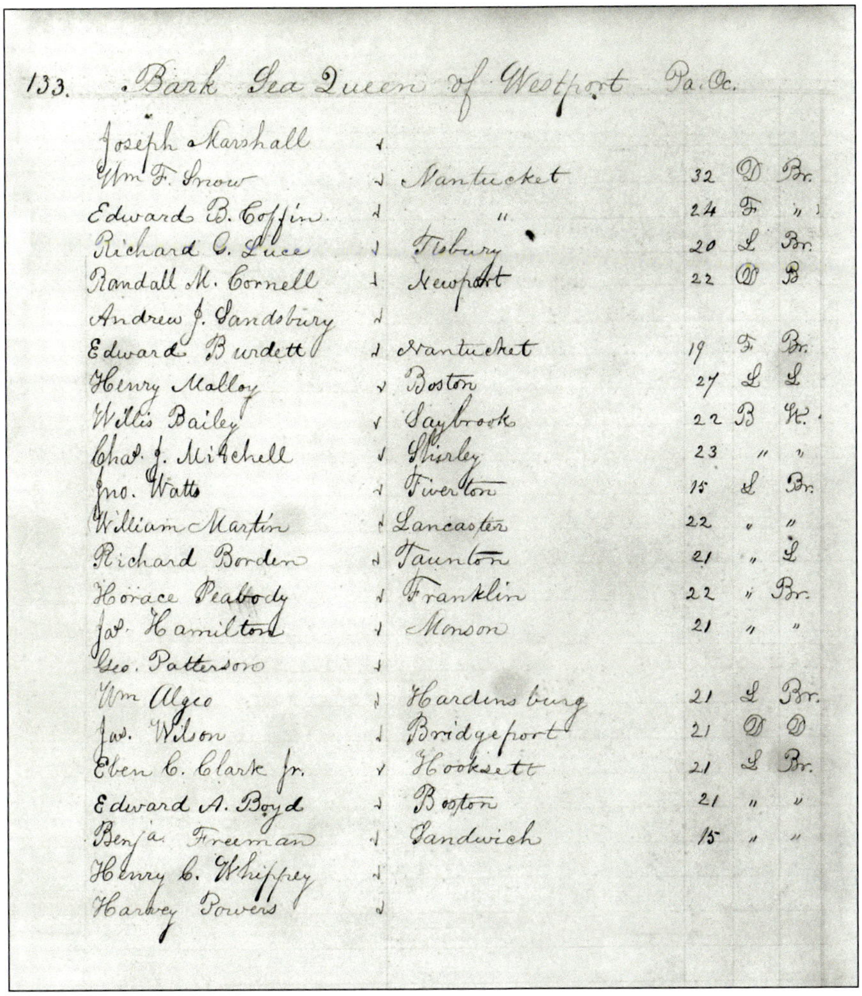

133. Bark Sea Queen of Westport Pa. Oc.

Name	Place	Age		
Joseph Marshall				
Wm F. Snow	Nantucket	32	D	Br.
Edward B. Coffin	"	24	F	"
Richard C. Luce	Tisbury	20	L	Br.
Randall M. Cornell	Newport	22	D	B
Andrew J. Sandsbury				
Edward Burdett	Nantucket	19	F	Br.
Henry Malloy	Boston	27	L	L
Willis Bailey	Saybrook	22	B	W.
Chas. J. Mitchell	Shirley	23	"	"
Jno. Watts	Tiverton	15	L	Br.
William Martin	Lancaster	22	"	"
Richard Borden	Taunton	21	"	L
Horace Peabody	Franklin	22	"	Br.
Jas. Hamilton	Monson	21	"	"
Geo. Patterson				
Wm Algeo	Hardensburg	21	L	Br.
Jas. Wilson	Bridgeport	21	D	D
Eben C. Clark Jr.	Hooksett	21	L	Br.
Edward A. Boyd	Boston	21	"	"
Benja. Freeman	Sandwich	15	"	"
Henry C. Whippey				
Harvey Powers				

Crew list from the logbook of the *Sea Queen*, 1851. *The crew of 24 includes men from throughout New England and beyond, including the captain and first mate at the top of the list. The two columns on the right abbreviate the men's skin complexion and hair color. It includes two black men, Bailey and Mitchell, whose hair color is denoted as "W," for "wooly."* – New Bedford Free Public Library

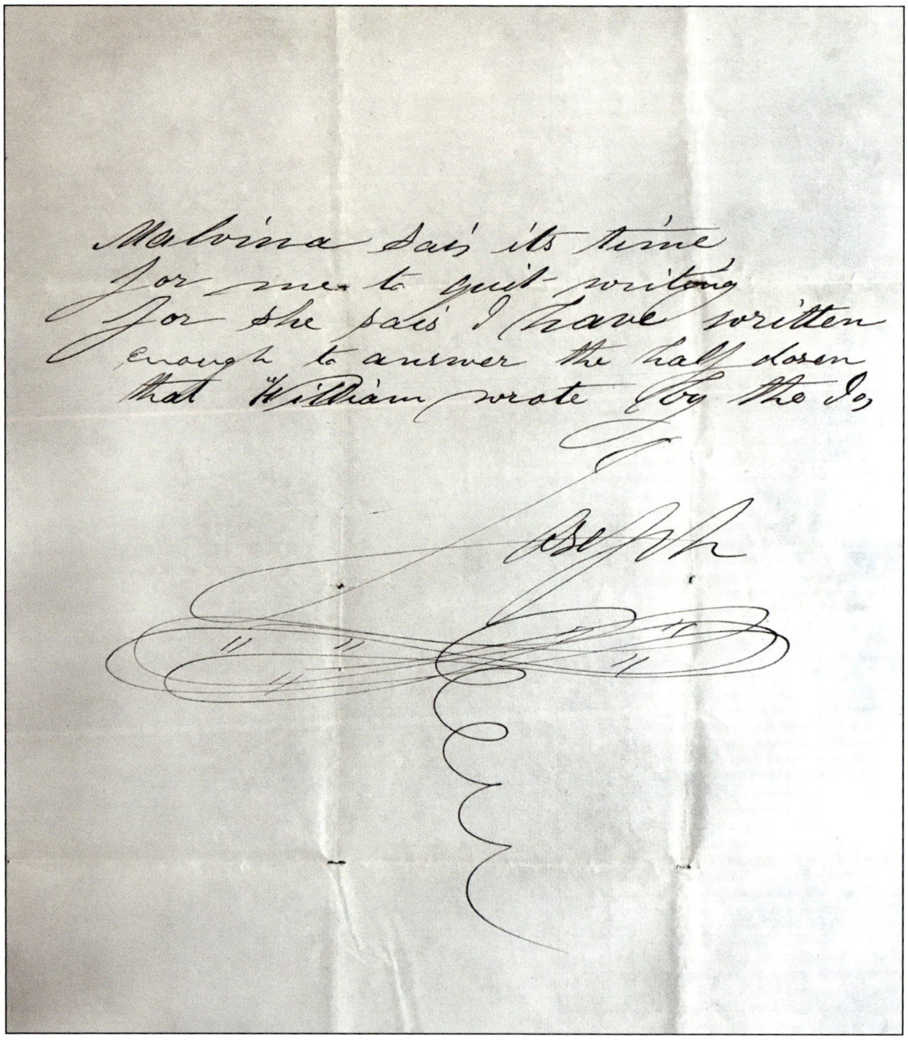

Malvina sais its time
for me to quit writing
for she sais I have written
enough to answer the half dozen
that William wrote by the Jo,

Joseph

A note from Captain Joseph Marshall to William and Elizabeth Crosby.

Typical of the Captain's letters and journals was the use of his distinctive signature. – Nantucket Historical Association

Chapter Two

The House with the Pink Door

Under clear skies, blessed with mild winds, and in sight of the approach to the Bay of Horta,[1] Captain Marshall put the third mate at the helm to take the *Sea Queen* into a safe anchorage. But ultimately, it was Mr. Coffin, the second mate, who was called upon to use his navigational and ship-handling expertise to maneuver the ship through the difficult channel to the inner harbor. Then, with anchor in place, he organized the lowering of the ship's boat and set several of the deck hands to work preparing the wicker gamming[2] chair and its four lowering lines.

Joseph helped Malvina up to the deck, for she had spent the previous weeks in extreme discomfort and was quite weak from acute nausea caused by the lack of fresh air in the cabin and by storms that continued buffeting the ship. Unable to take regular food, her diet had been limited to boiled potatoes and tea.

Aided by Mr. Coffin, Joseph secured Malvina to the chair using a wide belt strapped under the arms of the chair and fastened behind the chair-back. Then, with more speed than ceremony, he draped the ship's flag[3] around his wife for her formal entrance into Horta. Four crewmen guided the dangling chair into the ship's boat and rowed across the smooth waters of the bay to the landing at the quay. Requests for a sedan chair and a doctor had been sent to the hotel ahead of their arrival.

An unusually large gathering of local residents came out into the sunshine of the mild November day, while others leaned out their second-story windows to watch with curiosity as the procession from the *Sea Queen* made its way up the street and into the hotel.[4] Soon Joseph, Malvina, and Mr. Coffin were ushered into rooms that had been prepared with white linens and pitchers of cold drinking water.[5] While the baby was not due for another month, Malvina was quite sure that her confinement had begun, and it was with some concern that she learned that there was no doctor on the island.

So much needed to be done! Joseph made a hurried trip out to the *Sea Queen* to collect clothing and his medical guide,[6] which included directions for childbirth at sea. He ordered William Snow, the first mate, to take command of the ship for a month's cruise hunting for whales. Edward Coffin was left on shore to lease a house for the Marshall family, to oversee its furnishing, and to hire a cook and housekeeper. Then, he had to recruit additional crew and purchase a supply of food for the

Title page to *Mariner's Medical Guide.*
Captain Marshall used the 1848 edition of the guide. –Library of Congress

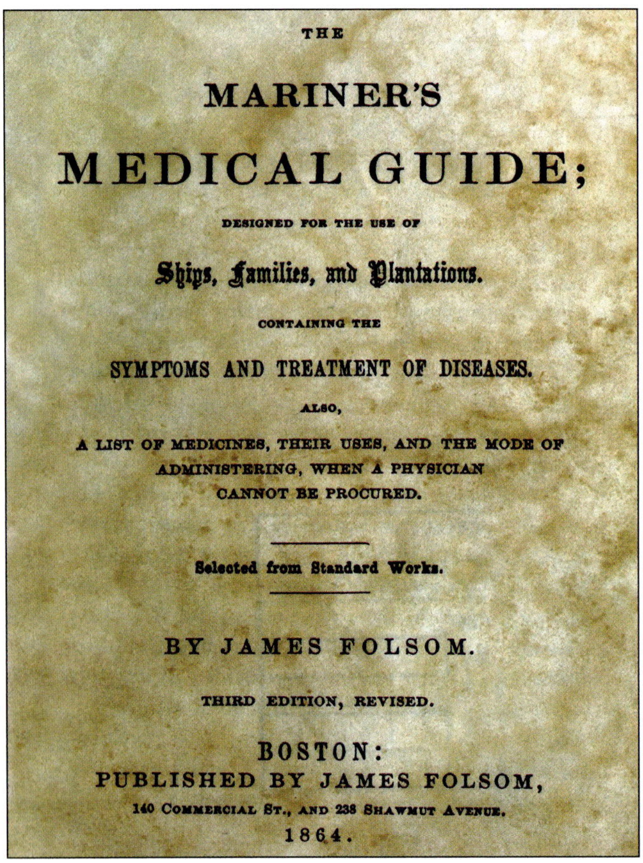
THE

MARINER'S

MEDICAL GUIDE;

DESIGNED FOR THE USE OF

Ships, Families, and Plantations.

CONTAINING THE

SYMPTOMS AND TREATMENT OF DISEASES.

ALSO,

A LIST OF MEDICINES, THEIR USES, AND THE MODE OF ADMINISTERING, WHEN A PHYSICIAN CANNOT BE PROCURED.

Selected from Standard Works.

BY JAMES FOLSOM.

THIRD EDITION, REVISED.

BOSTON:
PUBLISHED BY JAMES FOLSOM,
140 COMMERCIAL ST., AND 238 SHAWMUT AVENUE.
1864.

upcoming trip to the Pacific. Meanwhile, Malvina was thankful to be on land and to have Joseph at her side, but she was soon racked with a protracted labor exacerbated by an understandable exhaustion. It would be two full days later before the tiny, red, and wrinkled Helen Marshall was born on November 26, 1851.

The captain and the mate were as prepared as possible for the birth. According to traditions long-repeated by captains at sea, Joseph paid all his attention to his wife, delivered the child, tied off and cut the cord, and handed the infant off to the mate for bathing and wrapping. In the nautical culture, even the birth of a child proceeded according to the chain of command.[7]

With small but frequent servings of rice, pork, and milk specially prepared and delivered to the hotel by the family's new cook, Malvina soon began to gain strength. She was overjoyed with her baby daughter but became increasingly anxious to move into a proper house.[8] Finally, on December 3, the family of three moved into a two-story stone house gleaming with white-wash and a roof of bright red tiles. It faced a narrow street paved with basalt stone, no sidewalks, and a dooryard framed with pale pink roses, bright red geraniums, and shrubs of blue-tipped rosemary.[9] As Joseph and Malvina joyfully entered this pretty but temporary domicile, they could not help but picture their family home more than 2,500 nautical miles away—where winter had begun, where the gardens were dormant, and where Joseph's son remained in school, dreaming of when he could go to sea.

By tradition, Azorean families used the first floor of their homes for a shop or for storage, a practice that allowed for easy access to an outdoor kitchen protected by a portico. In the upstairs living area, the Marshalls' parlor had blue floors and bright flowered draperies adorned with long black fringes. The settee and chairs, while not as elegant as those of the Fair Street house on Nantucket, appeared sturdy and comfortable, and the large dining room featured a small pantry and a sideboard at one end. Next to the window stood a china cabinet that held enough tableware to serve the needs of a small family, and the table was covered with a heavily embroidered fringed tapestry.

One of the two bedrooms had a high-posted bed and an ample chest of drawers, with a small table and candle stand next to the bed, and Malvina's domed-top trunk at the opposite side of the room next to the armoire. But the other room was empty, except for Malvina's black Nantucket-made rocking chair, the baby's large oval basket outfitted with a thin mattress covered in pillow ticking, and a small chest or chiffarobe. One window provided pleasant light during the day, and a diminutive glass whale oil "courting lamp" shed just enough illumination for checking the baby at night.

In an alcove off the hallway was a small water closet with commode, wash basin, and pitcher that completed the necessities. Joseph and Malvina knew immediately how hard the mate had worked to provide for their comfort, but some things would require getting used to. Joseph would have to use space on the dining table for his temporary desk, while Malvina questioned the appropriateness of the large basket in place of a cradle.[10] Even more, as Unitarians only one generation removed from a Quaker family past, they were uncomfortable with the ornately carved and colorfully painted crucifix that was a permanent and prominent fixture in the home.

Helen's room in Horta. *Malvina's Nantucket-made rocking chair and Azorean-crafted child's basket-crib are in the collection of the Nantucket Historical Association. The whale oil glass lamp or "courting candle" is from the Bethny Brown Avila Collection.* – Drawing by Dawn Secord

Yet they were able to laugh when they put their tiny baby daughter in the middle of her new bed: Joseph suggested that Helen was the "size of a button left in the wash tub."

For Malvina, it was nice to have her Joseph back so completely, at least for a while. And although he couldn't neglect his maritime duties entirely during this time, they both were thankful for the privacy of a home.

To add to her contented spirit, Edward Coffin arrived with welcome surprises—the mail had just arrived from Boston on the packet *Io* and brought three letters from family and a package of *Inquirer* newspapers. Then their cook arrived to introduce herself—and after the housekeeper's many attempts to translate the spirited conversation from Portuguese to English and back again—Edward, Joseph, and Malvina were certain only that their first meal would be brought in soon.

After they had read the mail together, Joseph busied himself answering the family's letters so that replies could be delivered to the *Io* for its return trip. He addressed his first letter to William Crosby, husband of Malvina's sister Elizabeth:

> *Dear Brother, Elizabeth's letter has been received which was very pleasant to us indeed and Malvina will answer it as soon as she can sit long enough to write. She is very well except for a sore breast... the Babe is very well.... [Malvina] is anxious to get home with her as she has been to the World's Fair after her instead of [simply] sending to Boston for her.*[11] *Every day there have been dismasted vessels coming in. The wind is yet east and south-east and they tell me it has been blowing here from the eastward for over a month. And one gentleman who has lived here for 46 years says he never saw such a time. The Harbinger drove ashore and three others (including) a French brig. They came ashore and in thirty minutes there was nothing left of them. A Dutch brig arrived that was dismasted in the hurricane that we had. They lost everything from the deck and had only water soaked bread for 18 days. They rigged a jurymast and set a studding-sail and staysail to get along with. Their entire cargo of sugar was dissolved yet only two men were lost. I have sent a few oranges to you by the Bark Savannah, Captain Lunt. Should they arrive in good order please send Mother Marshall a half dozen. [Toward] the end of this month I shall go directly to the Pacific. Malvina says it is time for me to quit writing that I must have answered all the questions.*
>
> *Joseph*

He then continued with his work of making out shipment orders, contracts for new recruits, and arranging for the *Sea Queen's* inspection and repairs. Together they wrote a brief note to Joseph's ten-year-old son[12] (Joseph W. Marshall) informing him of his sister's birth. Then Malvina fed baby Helen and changed into a more proper dress for the midday meal. Tea was brought to the table with a plate of coarse white bread with a flour-dusted crust.[13] Then roasted pork and sweet potatoes were served together with fresh sliced oranges (the pride of Fayal) and a tall mug of goat's milk prescribed by the cook for the nursing mother.

After Edward left, Joseph and Malvina sat at the table to discuss travel plans. They both knew that it was unwise to take the baby on the continuing voyage to the Pacific while she was so small. And since Malvina was just beginning to regain her strength, it seemed that a longer stay in Horta would be beneficial to her health as well. Accordingly, it was decided that Joseph would take the *Sea Queen* out as soon as the ship was outfitted for the long trek to the Pacific, that Malvina would remain in Horta until the baby was strong enough to travel, and that mother and child would then take the *Io* to Boston and sail home to Nantucket until plans could be made for travel to Panama and Peru, where she would rejoin Joseph on his way to the distant whaling grounds.

For now, they would have but two weeks together.

Joseph had a compulsory meeting with the American consul, Charles Dabney, the next morning. Charles had succeeded his father, John Bass Dabney in that office, and among his responsibilities was to see that the local suppliers of goods and the recruits who signed to sail on whaleships were paid at a fair rate. He was not simply vigilant for the well-being of the island's people; he had become acculturated to the Azores and spoke fluent Portuguese. Originally from Boston, his family included highly successful businessmen who traded in whale oil, barrel staves, wine, and the myriad collection of local fruits, vegetables, poultry, and beef that was essential to maritime survival and commerce.

Charles William Dabney (1794–1871). *Dabney was U.S. Consul to the Azores in Fayal.* – New Bedford Whaling Museum

Captain Seth Pinkham of Nantucket (1789–1844) *was Malvina Pinkham Marshall's father.* – Nantucket Historical Association

As always, Joseph found Charles Dabney to be cordial, affable, and fair-minded, and on this occasion he learned that Charles had been a close acquaintance of Malvina's late father, Captain Seth Pinkham,[14] especially during the time that Pinkham had served in the Massachusetts Legislature at Boston. In addition to the business of the meeting, Joseph arranged to resolve a personal problem: the consul would send an English-speaking boy to help the Marshalls communicate with their house staff and their neighbors.

Accordingly, a few days later, Malvina woke from a nap to find the housekeeper poised to present a visitor. Paulo[15] was a slightly built fourteen-year-old with a "delightful smile and polite manners." And much to Malvina's surprise, he greeted her in English. After they talked about the Jesuit priest who had been his teacher, and the difficulty of Latin conjugations and declensions, they went to the baby's bedroom so that Malvina could gather up her fussing child. Paulo was much impressed by the fine Fayal basket these Americans had obtained for their child's crib.[16]

Now that they had a translator in the house, the Marshalls were able to find out the full story of their baby's bed. The cook's brother, who was witness to the family's arrival after the great storm, had woven the basket out of local reeds in a size and shape to keep the baby comfortable and safe on a ship at sea. Its high sides and ample length were meant to accommodate the child as she grew, because "it would be the ship that rocked this cradle."

When Malvina was at last able to go into the town for a walk, Joseph was happy to show her the marketplace, the churches and cathedral, and most delightful of all, the orange groves. She had never seen an orange tree before and marveled that the fruit and new blossoms grew at the same time. And the perfume of the trees mingling with the salt air was a new and glorious experience.

Her only concern was the often missing street signs and house numbers. With all the houses looking so similar, would she ever be able find her own home by herself? Joseph calmed her with a smile and said, "Our house has a pink door!"

Several days later, Joseph had to inspect repairs being made to the *Sea Queen*. Malvina left baby Helen with the cook and housekeeper and walked with him to the shoreline. Looking back at the city, she was surprised to notice for the first time that Nantucket and Horta were decidedly similar in some ways.[17]

Both towns were characterized by their many bare hillocks and relative lack of trees; buildings were nestled close together beginning right at the shoreline. High up on the right border of each city stood a windmill. To be sure, the architecture of the churches was very different, but their spires similarly punctuated the skyline above the roofs of dwellings, markets, and official buildings.

And as at home, blue hydrangeas were everywhere, even in this less than summery month, providing contrast with other flowers and a monochrome accent to the ever-visible ocean.

Still, of course, Fayal and Nantucket exhibited many differences, most prominent being that of language. Portuguese was the native tongue, though visitors often relied on some Spanish. So it was with Joseph, who could conduct simple business arrangements in Spanish, but had little knowledge of Portuguese. For Malvina, who had studied only Latin and French, the most noticeable difference was that the people of Fayal spent most of their time in their houses, very seldom venturing around town to visit with neighbors; only on market days were they to be seen out and about.

Then too, the fashion of dress for both men and women stood in sharp contrast to the Quaker attire or more modern and decorative choices made by the people of Nantucket. Fashion in the Portuguese culture was dictated by wealth and social position. Men of the leisure class commonly wore formal clothing and their womenfolk were fond of attending parties, balls, and other social functions in elegant silk ensembles that included mantillas for head coverings.

Yet women of the less fortunate majority were clad in long dresses of modest material, shawls, aprons, and white cotton head scarves topped with a flat hat, usually of straw. It was also common to see women wearing over-large blue hoods with attached capes that looked more appropriate to the European middle ages than the nineteenth century. The various tradesmen who made deliveries, repaired buildings and streets, and drove the carts and wagons required by the island's agricultural and commercial economy wore short wool jackets, three-quarter-length britches and either conical cloth hats or broad brimmed flat ones. Men who commonly worked in the fields also wore long white coat-shaped tunics and leather sandals.

Before long, the work of repairing and supplying the *Sea Queen* was almost complete, a full crew had

Overview of Horta, Fayal, late nineteenth century. – From a lantern slide, Spinner Publications Archives

been contracted, and all the paperwork for travel was in order. Malvina, nicely settled in her new home, had found the cook and the housekeeper to be friendly, caring, and as meticulous in their cleanliness as any New England housewife. Paulo was an invaluable helper in the markets and dry goods stores and proved intuitive at helping her with the difficulties involved in attempting to speak Portuguese.

Joseph and Malvina were rapidly growing accustomed to this new life and marveled at how fast little Helen was growing. Yet the parting of their family was inevitable, and during the week before Christmas, Joseph and the *Sea Queen* sailed for South America and the broad Pacific.

Letters from Nantucket brought Malvina both joy and occasional feelings of melancholy. Her sisters, brother, and mother kept their news cheerful and sometimes even comical, but descriptions of holiday gatherings and celebrations with neighbors brought tears to her eyes. Malvina began to plan her trip home. She first set to work buying material for the new clothes she would design and make to fit her new slender figure. Then she "re-worked her best silks and made new bonnets to match" and wrote lists of the belongings she would need to take with her: the basket, the rocker, all her linens, gifts of tea, coffee, oranges for the family, and two trunks—one for her and a small one for Helen.

In January, a delivery boy brought a personal note from Roxana Dabney, daughter of the American consul.[18] Malvina found the paper and penmanship to be surpassed only by the sweetness of the message. It was addressed to Malvina Pinkham Marshall—certainly a nice personal touch in its use of her family name rather than her title of "Mrs." with her husband's name.

Roxana, it said, would like to call on her at two o'clock in the afternoon on January 13.

The sewing and lists were immediately set aside to prepare for a tea with this young and fashionable daughter of Charles Dabney. Malvina would need to help with the preparations, since the cook had never heard of hearts and rounds or pound cake or lemon ice.[19] She extricated the shaped tin cutters for the little cakes from the bottom of her trunk, then set about the task of obtaining the tools to shave the ice and finding a way to keep it cold.

Roxana Lewis Dabney (1827–1913). *Daughter of Charles William Dabney, Roxana was born and grew up on Fayal, Azores, and was educated by tutors. She later attended schools in New England and is the author of* The Dabneys: A Boston Family in the Azores. –New Bedford Whaling Museum

The visit went exceedingly well. Both women belonged to the culture of the Bay State, for Roxana had spent some of her formative years in Boston and was well-read and an accomplished musician. They both shared a fondness for reading Byron and had long been accustomed to a regimen of exercise and family sports and games. Roxana told Malvina about her family's visit to the World's Fair at the Crystal Palace in London and their planned trip to Boston in April. She and Malvina would soon be traveling the same route and on the same boat, though on different schedules.

With the visit a delightful success and a new acquaintance made, Malvina turned to the future. Although tempted to wait until Spring and make the trip with the Dabney family on the packet, Malvina decided to book passage in February along with Paulo,[20] who would help with Helen and the baggage. And so it was that on February 14, 1852, Malvina looked back only briefly at her provisional home with the pink door, and with baby Helen in her arms and Paulo at her side, boarded the *Io* in Horta Harbor bound for Boston, with Nantucket as the final destination.

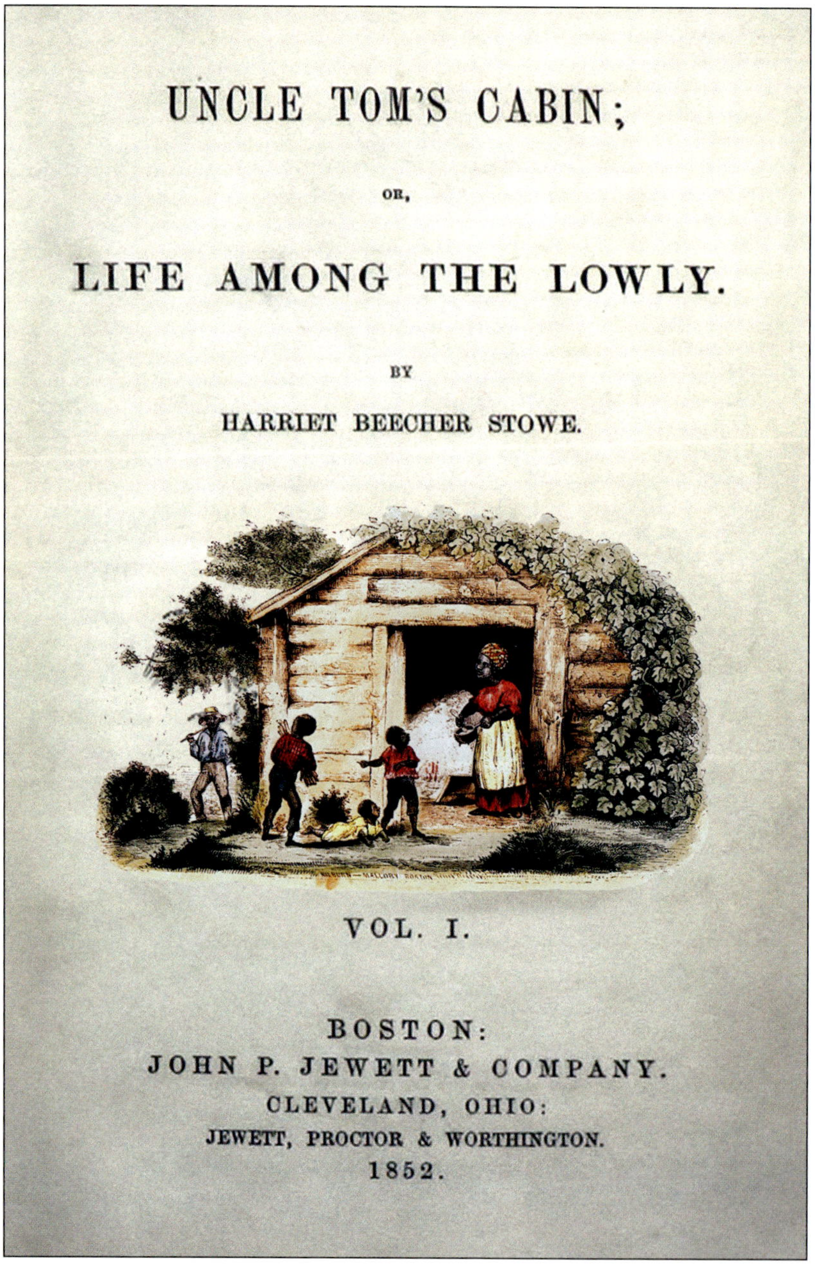

UNCLE TOM'S CABIN;

OR,

LIFE AMONG THE LOWLY.

BY

HARRIET BEECHER STOWE.

VOL. I.

BOSTON:
JOHN P. JEWETT & COMPANY.
CLEVELAND, OHIO:
JEWETT, PROCTOR & WORTHINGTON.
1852.

Composite title page from *Uncle Tom's Cabin*. *The classic, 1852 novel arrived on Nantucket the same week Malvina returned from Horta, Azores.* – Library of Congress

Chapter Three

Nantucket: Reformers and Rhetoricians

It was late March, at the tail end of the coldest winter yet recorded, when Malvina's brother, Seth Jr., met the *Io* in Boston, transferred the trunks and crated belongings to the wharf where the Nantucket boat would tie up, and arranged for Paulo's passage back to his family in Horta. He had brought extra blankets and oilskins to the dock to protect his sister Malvina and baby Helen from the severe cold and the impending inclement weather on the boat trip to their island home. Nantucket was only seventy nautical miles away, but floating ice in Boston Harbor could make sailing difficult, to say nothing of the harsh winds across Buzzards Bay and Nantucket Sound. Nevertheless, any idea of delaying the departure was dismissed at the thought of the welcoming party that family and friends would surely have waiting for them at home. Malvina was returning not only to "the family hearth," but also to a world that was soon to undergo major social, political, and economic upheavals. Among the first omens of this was a recent turmoil in the community that had been initiated by social reformers, had grown steadily for a number of years, and had recently been escalated by political leaders and orators of national acclaim.[1]

Nantucket had always been a community that practiced independence and self-reliance in thought and deed, and its early Quaker values and disciplines had sustained the people through frequent family separation, hardship, and danger. Their inherent commitment to honesty, thrift, industry, and simplicity were so ingrained that even when their children and grandchildren no longer "went to meeting," they were Quaker "in thought if not in practice."

This Quaker influence was also significant in establishing the expectation of female proficiency in many different endeavors. In fact, the roles and responsibilities of women in this seafaring society gave rise to a degree of gender equality that was seldom if ever present elsewhere.

Nantucket women managed the family finances, owned and sold property, traded in commodities, and ran their own businesses. Men relied on women to take leadership roles in the community, in places of worship, and to educate themselves and their children. Fathers placed great value on both sons and daughters who excelled in mathematics, literature, and foreign languages. They even arranged business apprenticeships for their daughters so that they would be financially independent and able to earn a living should the need arise. Husbands and wives sought to create partnerships of mutual interests, abilities, service, and work.

So it was that both men and women in Nantucket were often active in the reform movements taking shape during the first half of the nineteenth century. Many of them were ministers, teachers, and women who asserted their freedom from the gender constraints of speaking in public.

Lucretia Coffin Mott (1793–1880). *The Nantucket-born daughter of a whaling captain, Mott became a Quaker minister and a leader in the anti-slavery and women's suffrage movements. She was a mentor to Susan B. Anthony and Elizabeth Cady Stanton.* –Nantucket Historical Association

Lucretia Coffin Mott,[2] born on Nantucket in 1793, was one of America's first and most influential reformers. With a skill and confidence honed at the Quaker meeting where she became a minister, Coffin Mott capably employed formal logic, deductive reasoning, and examples from the scriptures and the classical philosophers to make her case for women's rights and abolition. She was the first president of the American Equal Rights Association, represented the United States at the Abolitionist Conference in London, organized anti-slavery societies, and as a mentor to Elizabeth Cady Stanton and Susan B. Anthony, became the prime mover of the 1848 Seneca Falls Convention on women's rights. She supported and addressed reform movements throughout the country, joining her views and commitments with those of her husband, James Mott, an equally assertive champion of abolition and suffrage. Yet, at the same time, she "lived in a sphere of domesticity with their six children" in a large home in Philadelphia.

As the Motts rose to national recognition, others also took up the causes they espoused. In fact, twenty years before the start of the American Civil War, Nantucket was already the setting for a significant reform event. Anna Gardner had initiated the first Nantucket Anti-Slavery Convention, drawing record crowds to hear such famous orators as William Lloyd Garrison and Wendell Phillips. And it was at that same convention that the recently escaped slave, Frederick Douglass, after initial hesitation, made his first public plea for freedom.

Since coming to the North, Douglass had learned that the men on the wharfs of New Bedford and Nantucket were often connected to job opportunities for black men in the whale fishery and might well have connections with the underground railroad. While this assumption of trust in the whaling community was not always correct, it provided Douglass with a considerable degree of safety. And to be sure, many Nantucketers—both whites and free blacks—were active in the underground railroad, hiding fugitive slaves in attics and arranging for transportation farther north. They taught in the black schools, were instrumental in passing laws providing for the schools, and in 1847, won the long and difficult struggle for school integration.

Another fervent abolitionist with ties to Nantucket was Levi Coffin Jr., a distant cousin of Lucretia Coffin Mott and whose parents hailed from the island. An activist with the Free Produce Movement who believed that slavery could be eliminated by boycotting all products made with slave labor, he proposed a ban on the major commodities that supported the economy of the slave states: cotton, rice, and sugar. Members of the movement were urged to renounce the purchase of these goods, to persuade others to follow their lead, and to promote the use of available substitutes. This became a popular movement on Nantucket and led to the establishment of a silk industry. Even when the Atlantic Silk Company closed in 1844 after ten years of operation, many people of conscience continued to favor the use of silk for clothing and home decoration.

In addition to the local speakers who found their way to every podium and venue on the island, many off-islanders were invited to share their messages. Frederick Douglass—who had become the leading reform speaker in America, commonly drawing crowds of unprecedented numbers—marked his third return engagement on Nantucket by reprising the 1850 oration he had originally presented at Boston's Faneuil Hall. Famous men and women arrived from the Chautauqua Circuit, and this in turn encouraged the formation of home-grown literary societies and debating clubs.

Even notable speakers who did not visit the island were accorded detailed coverage by Nantucket newspapers. These included Sojourner Truth, a former slave who was moving audiences to tears elsewhere in New England and the Middle Atlantic states with her account of losing her children at a slave market and her haunting challenge, "Ain't I A Woman?"

Still, Nantucketers were far from unanimous on abolition and racial equality. And although no extant letters document Joseph and Malvina Marshall's attitudes on these explosive social issues, we know that both had lived and worked respectfully alongside African Americans. Nantucket was a small island with a growing black population that was continually increasing with the arrival of former slaves and immigrants from the Cape Verde Islands. Joseph had served with black crewmen and had hired black stewards and crewmen on his own ships. Malvina grew up in a household where Treloney Pompey, [3] a resident of Nantucket's black settlement, called "New Guinea," was in command of the family kitchen. In fact, it was the "strong but wispy little Mrs. Pompey" who instructed the Pinkham girls on the correct way of clearing plates from the table and the appropriate method of rinsing the dishes before washing them. The Marshalls' experiences, as well as their adherence to the Unitarian principles of tolerance, their choice of newspapers and books, and their family's liberalism, would suggest that they were supporters of abolition. Be that as it may, the growing interest in social reform and political action, along with the founding of a number of other Protestant congregations on Nantucket, led to a considerable increase in public debate.

The often-heated rhetoric of ministers and representatives of the press, anti-slavery societies, and social reform movements stirred up the people of Nantucket and created emotional and philosophical antagonisms that sometimes split families. As early as 1842, this once uniformly pacifist community had erupted in riot when Stephen Symonds Foster delivered an incendiary speech in the Atheneum, accusing the island's religious leaders of being a pro-slavery "Brotherhood of Thieves."

By the time Malvina and Joseph Marshall had left for the Azores in 1851, they, like their fellow islanders, had long been aware of the local and national advocacy for social change. Yet in 1852, ten years after the firebrand Mr. Foster had raised riot in Nantucket town, Malvina was returning home with one goal: to recuperate among those near and dear to her. There would be walks in the snow, visits to friends in town, tea and whist parties, daytrips in search of the first spring flowers, and as the days warmed, gatherings and promenades on the beach. And sure enough, Malvina did partake in all of these diversions, but they were punctuated by serious conversations about race and equality and slavery issues that had escalated significantly.

Indeed, it was during the very week of Malvina's return that the first shipments of the two-volume *Uncle Tom's Cabin*[4]—a book that "...became the match that lit the fires of change" and altered the social and legal mores of the future—arrived on the island. The rhetoric of reform in the newspapers and from the podium thus far had been enlightening, persuasive, and even shocking, but this story by Harriet Beecher Stowe, written in conversational language about a black family split apart and fleeing from the evils of slavery, soon engaged the hearts and minds of America. It forced everyone to take a stand for or against the moral and legal right of one human being to own another, and would prove to be a precursor to the war already dark on the horizon—a conflict of such far-reaching consequences that it would eventually bring down the curtain on Nantucket's vital whaling industry.

Poster advertising the California Gold Rush, 1849. *The packet ship* Josephine *arrived in San Francisco in November of 1849 along with the Nantucket-based* Astor *and four ships from New Bedford. By late December, the* Nantucket Inquirer *reported that 592 had left the island for California.*[1] *The* Josephine *made a second voyage from New Bedford to San Francisco on January 11, 1850. She was accompanied by at least two other former whaling ships: the* John Adams *and the* Hibernia. –Spinner Publications Archives

Chapter Four

Gold Fever, the Isthmus, and a Family United

Not only were the issues of slavery, abolition, race, and women's equality major agents of social change in mid-nineteenth-century Nantucket, and the threat of war a growing reality, but the news of a major gold strike in California was a pivotal event for everyone, especially for those involved in the whaling industry. Between 1848 and 1849, five hundred men left the island in search of California gold, and forty-two whaling ships from the local area were refitted to carry freight and passengers to San Francisco.[1] In addition, new steamships were built to carry gold seekers from east coast cities to the Central American ports of Chagres and Aspinwall (Colon) on the Caribbean, where they began the short but arduous journey across the Isthmus of Panama to the Pacific Ocean.

And in the summer of 1852, the worldwide frenzy created by the California Gold Rush was a serious consideration for Malvina as she planned her trip to Peru. Both land and sea routes from Atlantic coast cities to San Francisco were swarming with travelers of all kinds, and most of them male. The steamers to Central America were often double-booked with passengers, and the ships that took the long route around Cape Horn were equally crowded and took twice as long. Those who traveled to the Isthmus via the city of Chagres faced not only dangerous weather, but also perilous tidal conditions where the Chagres River emptied into the Caribbean, and quite a few died while transferring from their ships to the small boats that would take them up the river.

In addition, travel across the Isthmus was so inundated with people that food, water, and lodgings were scarce and of poor quality. Joseph, being well informed about the routes and the relative safety of each possibility, sent Malvina careful directions for her bookings and instructed her to make arrangements for a traveling companion for the Isthmus crossing. He also wrote to his brother-in-law, William Crosby, to give Malvina any assistance she might need in planning her travel.

The many connections and means of transportation looked forbidding: first she must sail from Nantucket to New Bedford, then go from New Bedford to Fall River by train, from Fall River to New York on the Steamship *Empire State*, and from New York to Aspinwall, Panama, on the Steamship *United States*. After this, Malvina and baby Helen would take the Panama Railroad to Gatun, and from there travel by dugout canoes called bungos up the Chagres River, past Gorgona to Cruces, where they would rent mules for the trip over the rocky and challenging Las Cruces Trail to Panama City. Finally, a British steamship would take them south to Payta, Peru.

Though this was not the only way Malvina could reunite her family, it was by far the shortest and safest route to the Pacific, and she decided to follow the example of those Nantucket women she knew who had lived their lives to the fullest even in the face of danger. So in the first week of July, with tears in her eyes and some fear in her heart, she said goodbye again to her loved ones on Nantucket and set off on her trip to Peru.[2] She held up her daughter and looked at her almost with surprise; the child was now seven months old and had a new name given by her adoring aunties: she was now called Nelly, Nelly Marshall.

The weather was hot and humid when Malvina and Nelly reached New Bedford after a routine sail across the sound and the bay from Nantucket. Their trunks and hand luggage had been packed with care, separate from the crated household furnishings that would be shipped by other means directly to the port of Valparaiso, Chile. Malvina climbed up the steep steps of the Fall River railroad car and reached out to retrieve her daughter from the assisting porter. As soon as the cars began their repetitious clickety-clack, Nelly went to sleep and had a fine nap until it was time to board the steamer. Referred to as the "Mammoth Palace Steamer" in the company's euphoric handbills,[3] the *Empire State* was a "lavishly appointed side-wheeler." It was 230 feet long with three masts, was lit with oil lamps, had a mahogany-paneled salon, a "grand table d'hôtel," and multiple decks with "wide balustraded staircases."

Steamship *Empire State*, built in 1850. *The* Empire State *traveled the Fall River Line carrying passengers, freight, and mail to and from Fall River and New York City between 1850 and 1856.* –Illustration from J. B. Lippincott Company

Malvina was pleased with her room:

Although quite high since the house is nearly full, the carpet (in the room) would suit Mother as it is a Brussells with small figures of green, red and white which makes it very bright and cheerful looking. I have a pitcher of ice water standing on the table which to me is very great luxury. And when I went down to supper a most sumptuous meal was laid out—a feast to the eyes and the taste. Now that I have got along so nicely as far as here I feel if I had not come I should have been a goose.

After enjoying all the attention of family and friends at home, Nelly now had only her mother to entertain her aboard the ship. She most enjoyed sitting in Malvina's lap and playing the singing and clapping game of "Uncle John is Sick in Bed," making it difficult for Malvina to write letters describing her trip.[4]

The next morning, they would transfer to the *United States*. Walter Folger, a distant cousin, had already arrived in New York and managed to find Malvina and Nelly in the crowd and help them aboard. On instructions from Joseph, William Crosby paid for Walter's ticket to Panama City in return for his care of mother and daughter on the trip.

The two-week voyage down the Atlantic coast gave Malvina plenty of time to reflect on her Nantucket visit. First, she wrote home in great detail, thanking everyone—Elizabeth, Harriet, Rebecca, Helen, Seth, and Mother—for their love and kindness during her visit.[5] As she wrote, her thoughts wandered as she visualized each room, each house, and each occasion. She happily recalled her father's initialed pair of green leather fire buckets that still hung in a prominent place in the front hall and her mother's luxury in the kitchen—a large, modern cooking-stove. Again and again, she pictured the house at 46 Fair Street (today number 40),[6] for she had been only seven when the Pinkhams moved in from next door.

Malvina also found pleasure in remembering the day in June she and Rebecca and Helen took tea on the back porch at Sans Souci in Siasconset, and the afternoon she spent at Elizabeth and William's home at One Pleasant Street, where little Nelly was happy to lie on her blanket and stare up at the exquisite chandelier that glistened in the sunlight. Then she remembered watching her mother prepare loaves of graham bread and writing down the recipe as she looked on.[7] And finally, she struggled to arrange in order the navigational methods and computations shared by close family friend Maria Mitchell (pronounced ma-rye-ah), who was already a world-famous astronomer. This data would be important for Malvina to put to use later on the *Sea Queen*.

U.S. Mail Steamship Company's terminal in Aspinwall, circa 1854. *The mail cars bear the names "A.M. Hinkley's Passenger and Baggage Express" and "Specie U.S. Mails." Along the tracks are mail boxes labeled "Wells Fargo," "Adams & Cos. Express," and "Runnels Express."* –Lithograph by Charles Parsons from a sketch by F. N. Otis

The excellent service by the steamship crew made life pleasant for the travelers, and it was not long before they landed in Aspinwall and took a carriage to the new brick train station. On the morning of July 21, Malvina, Nelly, and Walter joined a group of thirty in a car with cane seats and backs, while two hundred more passengers pushed and shoved into the other small, open-sided cars, attempting, sometimes in vain, to find places to sit. Those who could not find seats had to pay for the right to trudge through a trail of stubble, mud, insects, and reptiles in a trackside strip beside the rails. When the train stopped at midday, the travelers took lunch at a halfway house before continuing the trip to the small village of Gatun, where the rail line ended on a promontory jutting out into a wide bend of the Chagres River. While even the slow speed of the train had provided some breeze and a respite from the tropical heat, the passengers found conditions extremely sultry when, at two o'clock that afternoon, they boarded the flat-bottomed dugouts that would carry them against the current up the river toward Gorgona. Since the cloth and palm frond coverings rigged on the bungos failed to provide sufficient shade for everyone on board, Malvina was given a large heliconia leaf with which to shade and fan Nelly, who was riding in a small basket in the bottom of the boat. The party of thirty rode in five boats, each of which was manned by four native boatmen who walked barefoot along a ledge around the top of the sides of the boat, from which they propelled and steered the boat with long punting poles. The scenery was beautiful and far different from anything Malvina and Walter had ever seen. Giant ferns, bamboos, palm trees, and heliconia provided a lush, green background for the dazzling display of bright-orange birds-of-paradise, tropical shrubs, and flowers of every possible color. In places where the river was narrow, the vegetation grew up to a lofty canopy that blocked the sun and prevented most of the intermittent

"Incident on the Chagres River," by German-born artist Charles C. Nahl (1818–78). *Nahl is considered to be California's first significant artist. On his way to San Francisco in 1850, he traveled through Panama and made drawings of his journey. In 1867, he rendered this scene into oil on canvas. It depicts a flat-bottomed boat being ferried by natives up the Chagres River. While long-poling their way around obstacles through shallow water, an "incident" causes a man to topple overboard. Meanwhile, travelers take refuge beneath a canopy on the boat to avoid the burning sun.* – The Bancroft Library, UC Berkeley

rain from drenching the travelers. Because of the noisy traffic on the river, native birds seldom appeared during the day, but the jungle was alive with the calls of eagles, macaws, parrots, and toucans as well as the chattering and whistling of monkeys in the treetops.

Malvina had very little space and had to sit as best she could on top of her trunk, all the time watching, with some trepidation, at how high the water level was on the sides of the boat. She fed Nelly at intervals, sponged her face and arms with water to cool her off, and when it rained, helped Walter cover her with an oilcloth. The other people in the boat were sociable and even friendly. Among them were a very pleasant woman from New Bedford traveling with her husband, Captain William Swain, formerly of the ship *George Chaplain,"* and a quiet, bookish young man from Vermont who had left Dartmouth and was on his way to California.

Because the strong current slowed the little fleet's progress up the river, the passengers grew tired as twilight approached, but it wasn't until ten o'clock that night that the boatmen pulled the bungos onto shore and directed the group to the "hotel" for tea and lodging.

The Isthmus of Panama. *The broken line marks the travel route along the Chagres River taken by Malvina Marshall with baby Helen in 1852. After arriving at Aspinwall, they took the train to Gatun.* – Drawing by Dawn Secord, based on the 1848 map from the *Common-School Geography: Elementary Treatise* by D. M. Warren, Philadelphia

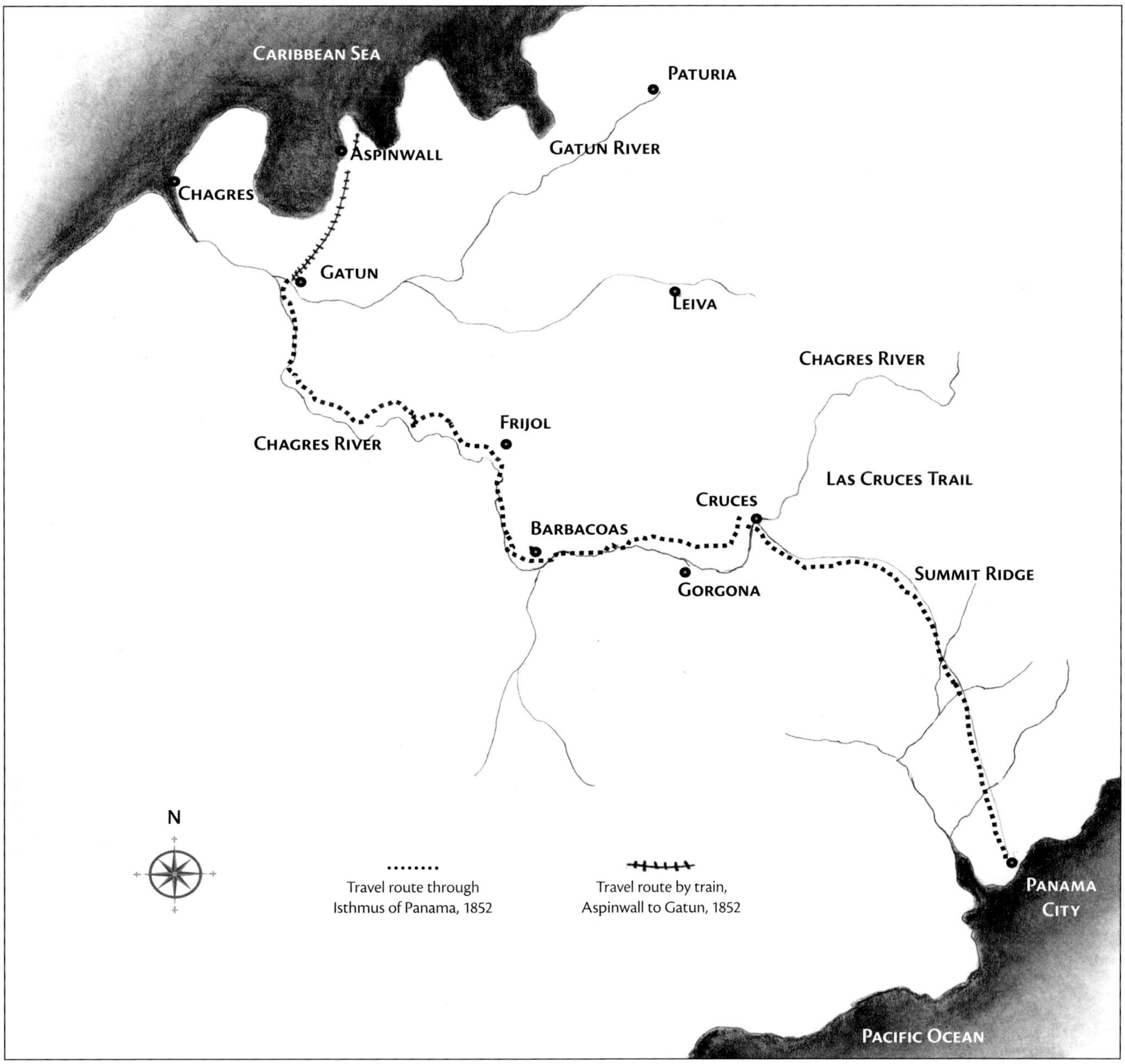

Malvina later recalled that she "had never been in such a hole before. Ants and roaches were galloping over the table." And though she was very hungry, there was little she could eat. The bits of meat smelled "turned," and even the plentiful supply of garlic and onions could not mask the bad taste. When it was time for coffee, the serving staff poured piles of sugar on the table to let the ants escape before it was spooned into the cups. As for the sleeping arrangements, Malvina later wrote,

> *The sleeping apartment for women was a large chamber on the second floor with ten cots put in, with only a blanket spread over them. There were no mattresses…For sleeping, I could not, for I watched the baby all night, so that the rats need not touch her as they were scampering round on the beams and having a nice time.*

Early in the morning, it was announced that the Gorgona Road was impassable from the rains, and all the passengers would have to board the boats again to continue up the Chagres River to Cruces. The heat was intense, and a dense mist rose from the teeming jungle in places where the river was bounded on either side by thickly matted tropical forests. It was not until noon that the group finally arrived in Cruces, a large village of some hundred huts now badly crowded with mules and impatient travelers from all over the globe. Each of the travelers was obliged to buy a mule that would be sold back to the company in Panama City, and they tried to make arrangements to depart from Cruces at dawn. Unfortunately, it was not until seven o'clock that saddles and harnessed baskets for the children and extra mules to carry the baggage were parceled out and all were ready to leave. Rather than proceed to Panama (City) via the Old Spanish Road, *El Camino Real*, the travelers mounted their mules and set out for the beginning of the Las Cruces Trail, which

Distant view of the old city of Panama, circa 1847. *Built on a peninsula reaching into the Pacific, the old city was little more than a fortress within a wall, and visited mostly by mariners, pirates, and conquistadors. It offered little more than trade in fishing or agriculure—until the Gold Rush of 1849.* – Spinner Publications Archives

would take them over the mountainous Summit Ridge to the Pacific.

Malvina would later write that it was the most stony and perilous road she had ever seen, and that no other animal but a mule could have carried the passengers safely over such a road. When the mules got their feet mired in some of the worst places, it seemed they would never get them out. Nonetheless, it was a fine day with a pleasant early shower, cooler temperatures, and a trail mostly shaded by the rainforest. In Malvina's words:

> *The ladies had to ride man-fashion, for the mules were not accustomed to side-saddles. [But] one lady in the party absolutely refused to ride that way and insisted she could ride sideways. Before the journey was over she had fallen off and been dragged from the mud so many times everybody's patience was at an end. The Indian guides took the babies of the party in baskets fastened on their backs, following the paths through the forest known only to them.*

When the guides rode off with her baby and those of the others, Malvina wondered whether anyone would ever see them again. But they soon enough returned, the babies were fed, put back into their baskets, and the riders disappeared once more. Malvina later recorded that "Nelly went over as safe and nice as could be. Her guide was clean, looked like Ben Clark, and was as pleasant and kind to her as I could have been."

Even though Malvina had been given the best and fastest of the mules, she did report a problem:

> *I came to a very bad steep place and my mule stopped and looked some time to find where to step. I let him alone knowing he would do better without urging. So I held on hard but he jumped and went another place equally bad so that before he could gain a sure footing he fell onto his knees and I went clear over his head onto the stones. I struck the side of my face and bruised it which made me a little faint. Walter ran to the rescue and was very much frightened, but as long as I was not entirely insensible I was not concerned. Walter and Mr. Nixon bathed my face with brandy and rinsed out the mud from my mouth. They took leaves from the trees and scrapped the mud from my bonnet and off my back and I was soon ready to mount again.*

Malvina's was not the only accident. Mrs. Nixon, a large woman from Iowa, was riding on a tall, skinny mule when her mount became stuck in a patch of soft clay. The woman was so heavy that the mule gyrated frantically in its struggle to get free, pitching Mrs. Nixon into the mud. But like Malvina, she was a hearty and resilient woman and insisted that the only thing to do was laugh and make the best of it.

The last three miles of the Las Cruces Trail was on a level road through open country, and being in a hurry to get to their journey's end, the mules trotted off at a fine pace. Arriving at the end of the trail an hour before the others, Malvina began to worry about Nelly. But all was well except for Malvina's appearance when she entered the hotel at six o'clock in the evening. She was covered with mud stains, her bonnet falling back over her bedraggled hair, her dress sash undone, her petticoat torn, and feeling so lame and tired that she had to be carried upstairs. Instead of taking one day to cross the Isthmus, it had taken all of Monday, Tuesday, and Wednesday.

Having done everything he could to help get his cousins to the Pacific, Walter left Panama on the *Winfield Scott* with inward satisfaction and in good health. Malvina was relieved that she and Nelly would soon be basking in the comfort of the British steamship to Payta, Peru,[8] but she still had to endure one more night in Panama and one more breakfast of rice cooked with strong onions and too much garlic.

Endure it she did, and indeed, the rest of the journey was far more pleasurable.

In a letter to her sister and brother-in-law, William Crosby, Malvina wrote:

> *As this end is the crowning of all things so has the last of my journey proved to be. I had the most delightful passage down from Panama on one of the finest steamships I ever saw. It being English, of course, we lived on board in the English style. My concerns about undertaking this trip have been at every turn put aside by my focus that it would bring such joy to Joseph. That joy is now real.*[9]

In the same envelope as Malvina's letter, the Crosbys received this message from Joseph:

> *July 28 – Much respected brother and sister, I have but a moment to spare but I must pen a few lines to inform you of Malvina's and Helen's arrival at Payta. I am going to sea as fast as possible only touched to get all that this world is worth living for.*

Map showing Atlantic and Pacific ports frequented by whaling ships. – Drawing by Dawn Secord

Chapter Five

Facing Pacific Horizons

On June 29, 1852, Joseph sighted the prominent and familiar headlands of Punta Payta, which sheltered the city of Payta from southerly winds and formed the western arc of the large crescent-shaped bay. This was his first attempt to meet the steamship from Panama and it ended in disappointment. The second trip was no more successful, but Joseph used his time wisely and managed to recruit four new deck hands, and purchase rice, pigs, and oranges. He also filed the ship's papers at the customs office and at the consul's office with Mr. Bathurst. There he learned from the loquacious consul that Malvina and Helen would arrive on the next steamship, and then posted a letter to William Crosby:[1]

> *Please send me at Payta, care of the consul, by the first opportunity: 12 dozen pint cans of milk, 4 dozen two pound cans of roast turkey, 4 dozen two pound cans of boiled chicken, 4 dozen pound cans of oysters, 1 dozen jars of pickles, 4 dozen two pound cans of roast beef from William H. Davis, no. 37 Broad Street, Boston Massachusetts. Those I brought with me were from Mr. Davis and were fine.*

His third landfall at Payta on July 28 brought success, and dressed in his best coat and formal top hat, Joseph went ashore in the ship's boat to gather up his family. On his return to the *Sea Queen,* he secured Malvina in the gamming chair, colorfully adorned with the ship's flag, and held his little daughter in his arms as they in turn were hoisted onto the main deck in a bosun's chair. Immediate preparations were made to "up anchor" and raise sails. The winds on the coast of northern Peru were hot and irregular, but they soon filled the sails, and the "*Sea Queen* made its way out into the Pacific looking like a speck on the broad ocean and surely walking the water like a thing of life."[2]

At last, after eight months of what seemed like a trip around the world, the Marshall family was united once more, and Joseph and Malvina were ready to face the vast Pacific and the shared efforts that would make their voyage a success. Since leaving the Azores, the *Sea Queen* had taken six blackfish[3] and no whales in the Atlantic, and only four whales in the Pacific, but the crew was now trained and eager to head for the whaling grounds west of Talcahuano, off the coast of Chile. All hands had worked hard to prepare the ship for Malvina's arrival, scrubbing the decks with holystones,[4] bleaching the quarterdeck, enlarging and painting the stateroom, and attaching a high-sided crib to the bulkhead across from the gimbal bed. And Malvina was as pleased at their thoughtfulness as she was impressed with the details of their labor. Now it was time for everyone on board to prepare for the strenuous business of whaling.

Malvina and Nelly were not the only newcomers on board. Recent recruitment at both Callao and Payta had brought on board seamen from the Sandwich Islands (Hawaii) and Peru. The captain and the officers were working hard to integrate the new men into the crew and were keeping close watch on the crew's ability to overcome language barriers and cultural misunderstandings. One way to pull the crew together was the use of boat drills. While sailing for the whaling grounds, Joseph scheduled a succession of drills, and by doing so, he was able to assess his recruits. He found the "Kanakas"[5] from the islands were fearless and "exceedingly strong at the oars," while the seamen from Peru followed orders quickly and thoroughly.

Malvina reacquainted herself with the areas of the ship that were within her boundaries and began creating activities that she and Nelly could enjoy in the sunshine of the quarterdeck. The furniture that had been shipped from Nantucket was not yet scheduled to be at Valparaiso for pickup, but Malvina had brought toys for Nelly to play with in the stateroom. Her favorites were a blue rubber ball that took on a life of its own with the rolls of the ship and a wind-up duck that quacked when it walked.

Remembering the slimy conditions of the deck during trying-out, the black smoke from the fuel that heated the try pots, and especially the smell from the processing of the whale that permeated every part of the ship, Malvina put her more formal clothes away under a layer of sheeting and arranged her clothes in the trunk so that the dark work clothes were on top and her oldest ankle boots were next to her full-length work apron. This way, all would be ready when the first whale was taken. This was the lifestyle[6] she had chosen when she decided to travel with Joseph, and such was the actual work of whaling, a job that had been part of their families' lives for as long as they could remember.

Even men who had dreamed of this "high adventure on the sea" when they were little more than children, quickly discovered upon first shipping out that this was a hard and dangerous life. They were often confronted with seasickness and then loneliness and worry about their families and friends back home, and not one among them was unaware of tragedy and loss. But when the man in the crow's nest cried "Thar she blooooows!," it would create a rush of excitement and alacrity of action beyond description. And the sailors on whaleships experienced it again and again.

On any given whaling voyage, as soon as a spout was spotted, orders would be sent to the helm to head the ship in that direction. Then, when the ship was as near as he deemed wise, the captain would luff his sails and give the booming command, "Lower away." The whaleboats would be swung out on the davits, each with an officer and boatsteerer aboard, and, typically, a thirty-foot double-ender would be carefully equipped with a tub of rope, harpoons, lances, a keg of water, a keg of hardtack, flint and steel, matches, and sometimes tobacco. As each boat was lowered, its four oarsmen would scramble over the side and into the boat, readying their long oars.[7] Then, as one slightly exaggerated account put it, "as soon as the boat touches the water she is off—sometimes to step a mast and hoist a sail." More often, the whale would not be far off, and they would row as near to it as they dared, then paddle softly to get close enough to dart a harpoon.

There were typically two harpoons ready to go, both ground as sharp as razors, and after the boatsteerer darted the first, he'd dart the second, whether to hit the whale or not, for the harpoon would be dangerous left swinging around in the boat. When the line began to pay out, the nearest man would pour water over it to keep it from catching fire as it ran over the gunwale. As soon as the boatsteerer darted the harpoons, he would scramble aft so that the mate might go to the bow, ready to dart a lance into the whale when the whale came up to blow.

It is the lance that killed the whale; the harpoon and line tethered the boat to the diving and often "running" whale. The men in the boat would have to hang on as best they could and keep the boat upright while racing over and through the waves at top speed in tow of a panic-stricken leviathan trying to escape. This "Nantucket Sleigh Ride" was perhaps the most dangerous aspect of whaling, since it often involved a choice of whether to risk the boat and crew, or to cut the line and let the extremely valuable quarry escape.

Because several ships could be hunting the same whale, the boat crew would mark their conquest with a waif—a white cotton flag with the ship's name. A chain would then be put around the flukes and the whale would be towed to the ship and made fast in preparation for the next two steps of the operation—the flensing, or cutting in, and the processing. But before those arduous tasks could begin, all the boats would return to the ship, and the captain would call the men to the quarterdeck to "splice the main brace." This ceremony, conducted by the captain, or in some cases the captain's wife, called for pouring each man a celebratory glass of grog.

Malvina later recalled what it was like on the *Sea Queen* after a busy day:

> *When the men had been hunting whales all day, the officers would "replay" all the actions over supper. They would stir their tea with noisy, clanking flourishes, pay no attention to what they were eating while describing with excitement how the boats were lowered, how close they got up to the whales, the sounds of the whales, and how long before the whale came up. Every detail of the hunt, the chase, and the kill—an old story to most but thrilling yet again.*

The next morning the demanding work of manufacturing oil would begin. A ship had to carry everything needed for this, and each man's skill at his job was essential to success. A brick try works was standard equipment on a whaleship and was erected on deck forward of the fore hatch. It consisted of a brick and mortar furnace containing two 250-gallon iron try pots supported by iron grate bars that allowed room for the wood fires on top of a brick and a wood foundation that was kept filled with ten to twelve inches of seawater to protect the deck from the heat of the fire. The fireboxes were held in place by the try work's masonry and iron sheathing. Secured on either side of the try works was a copper cooling tank and a spare try pot. The cooper or barrel maker placed his bench close to the tank and had his staves, hoops, and equipment close at hand. The process of cutting in a whale began with the carcass chained to the starboard side of the ship. The head was removed first, emptied

Deck plan and cross-section plan of the whaling bark *Alice Knowles*, 1878. *This vessel is similar in size and structure to the* Sea Queen. *Note the position of the bark's whaleboats, the three masts, and the try works.* –Sketch by C. S. Raleigh, from George Brown Goode

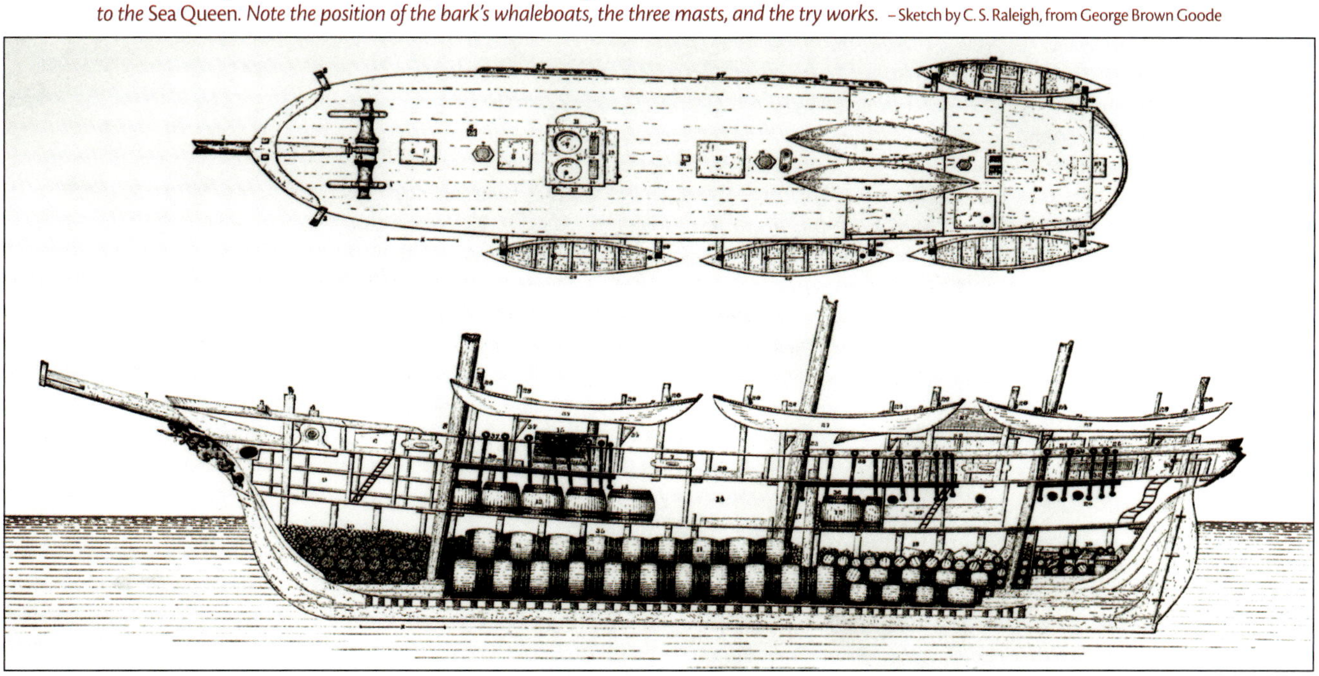

of its "liquid gold" (spermaceti), and if small enough, was taken aboard to remove the bones and teeth.

Then a boatsteerer would climb overboard onto the whale's back, and with almost superhuman efforts, fasten a large hook into a hole in the blubber cut by officers on the staging. Although he'd wear a belt or harness attached to a rope held by his mates, he faced grave danger from the slipperiness of the blubber and swarms of surrounding sharks. Before any further cutting was done, he would climb up the rope, often to a round of applause and an afternoon free from duties. Meanwhile, crewmen on the staging would use sharp spades on long handles to cut and peel huge strips of blubber from the whale carcass and hoist them aboard with blocks and tackle attached to the mainmast yards. Knives were used to cut the blubber into smaller strips, called blanket pieces, before feeding it into a mincing machine.[8]

With fires lit beneath the try works, the blubber would then be dumped into the try pots and boiled under careful supervision, for if the oil was boiled too much, it would not have a proper consistency, and if boiled too little, it would be "raw" and would not keep. Finally, the cooking would be done and the oil poured into the copper cooling tanks; its temperature tested in order to make sure it was ready to go into the barrels. Here the cooper's skill came into play, for the staves would shrink and the barrels would leak if the oil was too hot. And while all this was going on, the rest of the crew would have to man the ship, keep it on course, and try to keep the thick grey-black smoke blowing to leeward, for if the wind changed, the ship would be enveloped in the putrid, black cloud that arose from the boiling blubber and the fuel of wood and cooked whale scraps, or "cracklings." After the barrels were stowed in the hold, it would take days to thoroughly clean the ship, stow the equipment, and wash all the blubber and oil-soaked clothes by dragging them on lines in the ship's foaming wake.

With all this activity taking place on the *Sea Queen*, Joseph's responsibilities were many.[9] After a dead whale was brought back, it was essential that he keep the ship as stable as possible in order to keep the carcass chained to the side and allow the crew to do their work. Rough seas, dense fog, or driving rain often impeded the work of the whalers; in good weather and bad, however, it was up to the captain to maintain strict discipline for the good of the ship and the safety of the men. His orders were law, and the mates were careful to carry them out to the last detail, but punishments and dismissals were occasionally necessary. And even during the complicated and demanding work of processing oil, the captain had to always be planning how to sail the ship to the best whaling grounds, taking into consideration the time of year, the whales' migration routes, the availability of fresh food and water on the many islands, and often, the necessity of recruiting additional crewmen.

Like many whaling captains, Joseph had to take on a variety of other roles. Perhaps most important, he served as the ships doctor. When there were accidents—all too common in view of sharp cutting tools in daily use—he was in charge of cleaning and suturing the wounds. He also dispensed medications, set broken limbs, and occasionally played the role of counselor. Every Sabbath, he served as minister, choosing the hymn and reading the scripture lesson; in case of a death on board, it was he who officiated. Business functions also intruded on his time, for he was obliged to make regular reports to the ship's owners, file the required documents in each port of call, and occasionally make arrangements with homeward-bound captains to carry a goodly portion of his cargo of oil back to New Bedford, so as to keep hunting for more. In short, his job was much like that of the mayor of a small city, the community's doctor, leading businessman, commodities trader, lawyer, police chief, and religious leader all in one. And on this voyage, he had to find opportunities for being a husband and father as well.

Malvina's work was different, for there was no precedent, tradition, or job description for a captain's wife at sea.[10] To be sure, she had the full-time care of Nelly, and was strictly limited to the quarterdeck area of the main deck and the stateroom and cabin below. Since the cook was in charge of all meals and the steward served them at the captain's table, Malvina prepared food only on special occasions such as a holiday, a noteworthy birthday, or a social visit, or "gam," from other captains and their wives. In fact, Malvina was known to set the bar high when entertaining her "sailing sisters and their husbands" on the *Sea Queen* in the Pacific. In a letter to Elizabeth and William in October 1852, she reported that she and the steward had been busy arranging the rooms as neatly as possible, taking up the carpet coverings, dusting, wiping the glasses, and so on—just as they would at home if they were preparing for company. Her menu on this occasion was talked about until the story reached Nantucket and was described in the *Inquirer*: "Malvina Marshall of the bark *Sea Queen* [prepared and] served her guests warm yeast bread and butter, preserved blackberry conserve, cold boiled ham, plum cake, cheese, and [Brazilian] coffee." At a time and place where food and distinct ingredients were scarce or nonexistent, this bill of fare was impressive.

Malvina had other skills that were important to her family's daily life at sea. After she had graduated from the Coffin School, her well-to-do but practical father had seen to it that, like her sisters, she learned a trade so that she could support herself if the necessity arose. So, Malvina was trained under the expert instruction of Jerusha Gardner, mantua-maker and milliner, and although she had profoundly hated the entire year of work, she was thankful as an adult for the skills she brought to her role as wife and mother. She was very glad to be able to mend and alter garments as well as design, cut, and stitch new items of clothing for Joseph, herself, and their growing toddler. When she received newspapers and magazines from home, she was able to use the illustrations to create an up-to-date and fashionable bonnet for herself and a very small edition for Nelly, with an extra-wide poke to protect her from the sun.

To Malvina, cleanliness was next to godliness, and as Joseph's ship was always carefully organized and well maintained, she created jobs for herself to help meet those goals. When the cook gave permission to "enter his territory" and the ship was free from round-the-clock labor, Malvina would make soap. It was a long process, but with her mother's recipe and a hot stove, she was able to turn out enough bars to wash everything she had on her list with plenty left over to store for the future.

Then she would begin a project that had no end. For example, every whaler was amply populated with cockroaches as large as June bugs, and Malvina undertook the work of making the *Sea Queen* free of the pests. She and the steward toiled tirelessly at the task; first they poured "a scant cup of molasses" into each of a half dozen pails that were then filled three-quarters full with seawater and placed them in several 'tween-deck compartments. Then, according to Malvina, "morning after morning the steward would pour overboard pail after pail filled to the brim with cockroaches. There seemed to be as many thousands of them as ever." But neither Malvina nor the steward gave up the fight.

While a clear division of labor between men and women existed in most of American society in the middle of the nineteenth century, on the *Sea Queen* Joseph and Malvina worked together at two essential jobs. The first was mundane: the family laundry. Tubs were set out on deck when there was enough rainwater to wash clothes. Carrying the tubs from storage below, filling them with water, and then transferring the wet clothes to the rinse water required greater strength than Malvina possessed. So, Joseph did the heavy lifting and Malvina set her hand to the washing, using her own large bars of yellow soap. They worked together to hang the outerwear on deck and their personal apparel in the stateroom. The only laundry that Malvina did by herself was the daily two-bucket load of diapers, and when she hung out the long, white, rectangular diapers on the main deck, they looked to her "like ghosts dancing in the wind."

Their other cooperative endeavor was one vital to everyone aboard: navigation. Joseph was fortunate to have a new ship with all the latest equipment. A large compass hung in gimbals in clear view at the front of the wheel housing, and the quadrant and chronometer were carefully stored in the captain's cabin next to the maps, navigation charts, log tables, and the ship's logbook. Joseph was a licensed navigator who, as a master mariner, was permitted to sail an American ship of any size anywhere in the world. In addition to his experience of twenty-four years at sea and eight years as captain, he possessed an academic education superior to that of most other captains. He had continued his studies in geography, navigation, and mathematics on Nantucket until he went to sea at age seventeen—a late start for most men—and he even continued his studies in night school between voyages. And his advanced education had proved beneficial for his future success. Yet, despite his extensive knowledge and experience, he shared the important task of navigating the *Sea Queen* with Malvina.

> *Each day at noon the captain and the mate and Malvina 'took the sun.' that is, they found the altitude of the sun with the quadrant, counted the seconds on the chronometer, and with their book of logarithms, worked out the latitude and longitude for the day. In the cabin a chart was spread out on the table and the distance and direction were traced for the ship, showing exactly where [it] was.*[11]

Malvina learned to count the seconds accurately—not an easy task—and using the figures provided by Joseph and the mate, she computed the ship's position. The captain, mate, and Malvina would then compare their results and make any necessary changes before recording the data. Back on Nantucket, when astronomer Maria Mitchell heard of Malvina's work at sea, she immediately remembered her friend as a "superb mathematician skilled in the use of a sextant."

All this is not to say that things were always routine and methodical during the next three years of the *Sea Queen's* journey. There were many challenging Pacific horizons during that time:

> *"The* Sea Queen *took whales off the coast of Chile and, processed into 360 barrels, the shipment was then transferred in Valparaiso to other ships returning to New Bedford. Then they proceeded to Honolulu to follow the yearly whale migration south, and their 'greasy luck' kept all hands on board busy for over a month."*[12]

Nelly had her first birthday on board and then took her first steps. While the ship was docked in Callao, Malvina took the train to the bordering city of Lima to tour the Cathedral and shop at the European-style marketplace, and Joseph rewarded the crew for their hard work by buying forty-eight pounds of butter[13] for the sailors' larder. On the way back to Talcahuano, Joseph made a fortuitous stop at Chatham Island, now known as San Cristobal, in the Galapagos Islands, "where the men took on board 116 turpin"[14] [terrapins or land turtles] for fresh meat and filled the demijohns with water.

The whaling had been successful, and the ship was nearly full, but on arriving at Talcahuano, the *Sea Queen* was turned away due to a city-wide epidemic of small pox. Illnesses were always a source of fear among the ship's crew, and all were worried, except the sailors who had previously contracted small pox and were thus immunized or those who had been inoculated. But some on board were at risk, and Joseph and Malvina feared most of all for Nelly. Accordingly, the ship's course was set to avoid the heavily populated cities of the South American coast and to make one last swing around Pitcairn Island, where they would hunt whales and trade for yams, bananas, chickens, and water before heading for Cape Horn and home. Again, the decision was a lucky one—first they encountered blackfish and then more whales. The ship was soon full, and they stopped only for gams and an exchange of mail with the *President* and the *Charles* out of New Bedford.[15] And this mail was important. Because they could not post letters at Talcahuano, sending the mail with ships sailing north up the west coast of South America was the best possible way to get word to those in New Bedford and Nantucket concerning the *Sea Queen's* location and planned route of travel. With luck, the letters would get to Payta then travel the fastest route across the Isthmus to the Atlantic and connect with a steamship going north to the United States.

Nelly, now three-and-a-half years old, had begun to spend more and more time studying the ambrotypes and daguerreotypes of her Nantucket family so she could put proper names with faces. And she looked forward to finding someone on Nantucket to play with. Even before the *Sea Queen* prepared to round the Horn, Malvina began cleaning the stateroom and packing the family's personal belongings. She sorted the clothes, checking for damage from mildew, and then placed the heavy clothes on top. It would be cold on board as they sailed around Cape Horn, and even if it was going to be spring when they arrived in New Bedford, it could be cold there as well. She planned her wardrobe for arrival to look her best and calculated how tight she would need to draw her lacings for the new dress she had designed and made. Joseph was pleased with the success of the voyage; he was particularly appreciative of the contributions of his first mate, Mr. Snow. The officers and crew were jubilant over the fact that each man's "lay,"[16] or percentage of the profits, would be a heartening reward.

Although everyone on the ship was in high spirits about leaving for home, the old fears about rounding the Horn became real among the crew. And there was no way to estimate the time it would take for the trip home due to the great unknown of the weather and seas at Cape Horn. However, all concerns were immediately forgotten when a pod of whales was sighted off Juan Fernandez Island. The seas were calm, and the officers and crew took two whales with little difficulty. Working together at maximum efficiency, they processed and stored the oil by taking the barrels of slush—saleable fat drippings saved by the cook—and strapping them on the weather deck to make room for the additional barrels of oil. The extra time spent now required the *Sea Queen* to seek more provisions and water before sailing south. Joseph put into Santa Maria Island, a familiar stop for whalers, and besides taking on board a good supply of potatoes, rice, beans, and salted pork, filled the ship's water jugs and proceeded south for the Horn.

The temperature dropped, the winds increased, and without the sun, it became impossible to take sightings, but when they entered the Drake Passage, the winds shifted and came from the west, aiding their eastbound crossing. Joseph ordered extra lookouts lest the ship be blown off course toward the South Shetlands or stray too close to the headlands of the Cape. While the high seas at times posed a threat—especially to the barrels and equipment lashed on deck—the ship made steady headway and proceeded without incident. And so it was that two weeks later, when the sun at last appeared in the sky, the *Sea Queen* was just north of the Falklands and on a direct course for Pernambuco, Brazil. When the time came, Malvina, who was nearsighted, stationed herself next to the mate on the quarterdeck so she could celebrate "land sighted" from his first glimpse. And after dropping anchor in the giant international port filled with sailing ships from around the world, the *Sea Queen* lowered boats to transport both officers and crew in turns by watch to the city. And Malvina observed with singular interest as her little daughter struggled to "lose her sea legs and walk on terra firma."[17]

After five days in Brazil, and with the ship carefully inspected, everyone was anxious to sail at last for home. On the way, the crew worked hard to finish the hand-made crafts they were making and gather up the gifts they'd bought for loved ones waiting at home. Malvina made notes in her journal of events, which she wanted to share with her mother and sisters, and Joseph looked forward to a long layover with plenty of time to see his island anew through the eyes of his daughter, "to see the gas lights on Main Street, and to walk again among the pine trees."[18]

Watercolor of the whaling bark *Sea Queen, circa 1857.* *Built in 1851 at Dartmouth for* Andrew Hicks *of Westport,* Sea Queen *measured about 100 feet in length with a 24-foot beam (width).* – From the journal of the New Bedford bark *Chili,* Captain Benjamin S. Clark, 1856–1860, New Bedford Whaling Museum

View of Nantucket town from the Unitarian Church steeple, 1873. – *Scribner's Weekly*, Vol. 6, No. 4, August 1873

Chapter Six

Blest be the Tie that Binds

The steamer was to leave New Bedford at nine o'clock in the morning, and as the workers on the pier cast off the ship's lines, the mate blew a long, deep, ear-splitting blast to announce that it was under way. Nelly was startled by the sound that came from right over her head, but she quickly returned to peering intently from the deck's railing, where she watched as the gray sea chest was loaded onto a cart to be taken to the steamer's baggage room. It was her father's and had "J Marshall" stamped into the wood under the brass lock. She had also seen her mother's oval-topped trunk carried up the gangplank and just barely made out the small, green trunk with leather handles sitting on top of all sizes and shapes of luggage, boxes, and freight. That one was hers, and in it she had packed her new dress and her shiny new shoes, but she was above all concerned for the safety of her treasures. One sailor had made her a checkerboard and checkers with her initials on each piece, and another had given her a little doll, dressed like a pirate, as a farewell gift. Also in her trunk, along with the many-colored shells she had collected on the beach at Fiji, was her new book, *The Fox and the Geese,* sent to her at Payta by her brother, Joseph.

Captain Joseph Marshall's sea chest.[1]
– Elizabeth Marshall Steinbaugh Collection, photograph by Kate Singh

The steamship *Massachusetts* had a big, enclosed deck with lots of room for walking—or even running if Malvina would allow it—and she was excited when Joseph had told her that they would eat in the steamer's dining room. Because the trip to Nantucket would take all day, they had their own stateroom with a number on the door and a key to ensure privacy should they wish to sleep or read in quiet. But Nelly was curious about all the people on the boat. One family had a hairy, old, spotted Spaniel that she got to pat, another family had a songbird in a cage, and a group of women whose over-the-arm baskets were filled with yarn and long knitting needles sat on the padded benches in the lounge and commenced their busy-work with very little conversation between them.

Much to Nelly's surprise, a young man walked toward Joseph and Malvina, and with a broad smile and an excited "hello," asked if the little girl with them was Nelly. It was Lydia Brown's son, who now worked in the Nantucket post office. It seemed amazing to Malvina that although they were not quite home, they were seeing and being seen by folks they knew. The last hymn they had all sung together on the *Sea Queen* after dropping anchor off New Bedford was appropriate to that occasion, but now the words rang again in Malvina's ears: *Blest be the tie that binds.* She thought of family waiting for their return and of how thankful she and Joseph were to share a common Nantucket history—one that included family, friends, neighbors, events, and place of worship. When they were at sea and had quiet moments together, they often reminisced about their favorite places on the island—and now they could visit them again together.

The gray skies allowed for very little visibility, and it rained all the way to Martha's Vineyard. Then it rained very hard, and darkness began to set in during the final hours of the trip to Nantucket. By the time the steamer entered the harbor, Joseph and Malvina could barely see the muted lights of the town. They woke Nelly from her nap to dress her in her oil-cloth cape and hat. William Crosby drove his horse and carriage up close to the wharf "to collect those dear ones who were so precious to him." He and Joseph were much more than brothers-in-law; they were the closest of friends and trusted each other with all their business and family affairs. Naturally, this long-awaited reunion brought tears to everyone's eyes and hugs and kisses for Nelly.

Bumping along over the wet cobblestones that sparkled in the light of the gas lamps, they drove up Main Street, past the Rotch building, and with a slight jog to the left at the Pacific Bank, continued on Main to 46 Fair Street,[2] where their family was waiting. This house, built by Captain Seth Pinkham in 1828 for his wife Mary and their children, was neither grand nor imposing, but rather a sturdy and practical home for a growing family, "with two storeys and a half for the main portion, an ell of two storeys at the back, and a neatly balustraded 'walk' on top. Above the front steps, which were arranged in that Flemish manner…with a flight from opposite directions, was a canopy." The large parlor to the right of the front hall had double windows looking out to the street, and to the left was a slightly smaller sitting room with a single window, where the Pinkhams dined each evening. The house boasted several modern improvements: "a cooking-stove replaced baking kettles and tin 'kitchens' on a hearth, and a [long handled] pump in a little passageway behind the kitchen brought running water from a cistern in the yard." A large central chimney served a fireplace in each room, and a half-cellar and sub-cellar provided cool storage for preserves and dairy products.[3]

But most important of all, it was a household of "impromptu jollity, vivid personalities, individuality, and joy of music, while at the same time adhering to the rules of decorum and [proper] English speech." And for this homecoming, it offered ample space for all the aunts, uncles, cousins, and grandchildren from the Pinkham and Marshall families. For Malvina and Joseph number 46 was home. Home at last! Yet there were "empty places at the table." Malvina's oldest sister Mary and her husband, Captain Henry Plaskett of the China Trade, were on a clipper bound for Canton; Seth Jr. and his wife had moved to Boston; and son Joseph W. Marshall was sailing to Calcutta on the *Mercury*. But Mary Pinkham was the hostess for the occasion and Eunice Macy Marshall stood in equal regard as Captain Marshall's mother.

The families were linked not only by their close proximity on this small island, but also by their resident longevity on the island. Surnames of the founding fathers too often define the generations rather than taking equal note of the family's female lineage. The Marshalls' Nantucket roots went back to the early 1700s, when the first of their line, Joseph Marshall, married Abigail Hussey and had one son, Joseph (1720–1802). He married Phebe Folger in 1740 and they had thirteen children, the second of whom was Obed Marshall (1744–1817). Obed was a cordwainer—a designer and maker of shoes and boots from new leather—and became sufficiently successful to purchase six properties in town. He married Susanna Burnell and the couple named their son Obed (1787–1848) as well. This second Obed Marshall became a cooper or barrel maker, a trade indispensable to the whaling industry. He married Eunice Macy (1791–1874), and they and their eight children made their home at 3 Milk Street, next to the Court House on the corner of Main in the section of town called "Court End." Captain Joseph Marshall (1811–1879), the oldest of that family, married Malvina Fitz Alan Pinkham in 1846. Although the Marshall name is prominent in Nantucket history, the fact that Joseph Marshall's mother was a Macy bespeaks a lineage that traced back to the 1600s, to two of the first English families to settle on the island—the Macys and the Folgers. In that small, tightly knit community it was hardly surprising that the Pinkham family could claim similar roots.

Richard Pinkham III (1665–1718), the son of a carpenter, was born in New Hampshire and settled in Nantucket after marrying Tristram Coffin's granddaughter Mary in 1683. Their son Theophilus, who

The home of Seth and Mary Brown Pinkham, 46 Fair Street. *Built in 1828, its address is now identified as 40 Fair Street.* – Nantucket Historical Association

married Deborah Paddock and had ten children, was the father of Jethro Pinkham (1745–1820), the cobbler. With his ever-patient wife, Susanna, Jethro—probably with the help of an inheritance from his grandfather Richard—was able to purchase a sizable gray-shingled house "on a slight dip on Fair Street just before the cobbles decline onto Farmer." As birthright members of the Society of Friends and loyal Quakers, the Pinkhams avoided both external and internal furnishings that could be construed as worldly goods, nor would their meager budget have allowed for such things. They had eight children to feed and clothe and "take to meeting every First Day in their broad-brims and coal-scuttle bonnets." So, it was deemed that a little "triangle of grass at the door yard with a simple rail enclosure to protect a few currant and gooseberry bushes, some beds of mint, and a common little rose-tree" was quite enough. On the side of the house closest to Farmer Street was the addition of a "wart," or enclosed shed that contained the tools of the cobbler's trade, and on most days the cobbler himself. It was also a common meeting place for Jethro and his pals to swap yarns.

On Sunday, July 9, 1786, Seth—the youngest boy in the Pinkham family—was born "into this affectionate, rough and tumble household," which often faced hard times and poverty in those days. The Revolutionary War had decimated the whaling industry and often made it difficult to obtain essential supplies from the mainland. Even after England had declared Nantucket a neutral community, the work of rebuilding and fitting out whaling ships imposed overwhelming physical and financial challenges. And it affected everyone on Nantucket. The Pinkhams, like most other families, had to look for new ways to earn money. Susanna and the girls hired out as nurses for the sick, and they made cloth tapes, spun wool, and sewed quilts.

But tragedy was an ever-present reality. The ship *Alliance* had sailed home from Delagoa Bay, Mozambique, and entered the harbor with its flag at half-mast. Susanna's brother, Captain Bartlett Coffin had died at sea. And sadder still was the news that the family's fourth child, Reuben, mate of a ship before he was twenty, had died and been buried on the island of Java. Only the later receipt of Reuben's lay from the voyage could slightly ease the pain and temporarily relieve the financial burden for the family. Still, this payment for sea-going employment, however tragic its source, somehow instilled in young Seth a positive view of the future. He knew that a good living could be made at sea, but when still a child, he had to leave school to go to work in the Russell rope-walk, where they made Russian hemp into tarred rope and manila hemp into lines for ship's rigging and harpoon lines. In the summer, Seth crushed clamshells for walk-ways and picked berries to sell. He learned to drive heavy cartloads of oil and worked in the hayfields at harvest.

When he finally shipped out on a whaler at the age of fourteen for the Pacific under Captain Uriah Bunker, his dream became a reality, and he felt it his destiny to be ship's captain. Nine years later, having risen to first mate, with money in the bank, and the promise of a ship to command, Seth married Mary Brown, daughter of Captain William Brown and Elizabeth Coffin Brown. The well-educated Mary tutored her husband, helping him catch up on the studies he missed in his youth. Their seven children also benefited from their parents' passion for learning and came to appreciate their pedigree by studying the family chart that Seth painstakingly drew out for them. And while the Pinkhams were documented in town records, "Mary Brown's lineage lept (sic) off the page of island history." She was the granddaughter of Hezekiah Coffin, the Quaker captain of the *Beaver*—the ship of Boston Tea Party fame—and a direct descendant of island patriarch Tristram Coffin.

But the children of Seth and Mary Pinkham also took on diverse challenges in the lives they chose to lead. First-born Mary and fourth-born Malvina went to sea, whereas Elizabeth garnered the attention of Nantucket society and became the family archivist with her long, frequent, and detailed letters. She married William Crosby on Christmas Day in 1834, and together the very young couple—he only nineteen and she eighteen—began making plans to build a "mansion" at One Pleasant Street. After graduating from Moses Brown School in Providence, Rhode Island, William had joined his father in the oil and candle firm of Matthew Crosby & Son and inherited money from his extraordinarily wealthy grandfather, Zenas Coffin. Accordingly, land was purchased, old houses were torn down, and the construction began. The result was a stately Greek Revival edifice that was both fashionable and elegant:

> *It sits on a high basement—where the kitchen and dining room were located—and sports a handsome portico with two Doric columns atop a granite stoop and steps. The pedimented gable featuring a fan window faces the street, and a cupola sits on the ridge of the roof. The façade of the house is clapboard with pilasters separating the three front bays. Eight floor-to-ceiling first floor windows admit copious light to the formal double parlors on the north side of the house.*[4]

Elizabeth selected "silver door-knobs, French windows opening on a balcony of wrought iron, double parlors with folding-doors between, marble mantel-pieces, hand blocked wall paper, and the stateliest things in furnishings…" including a mahogany bedroom suite with hand-carved scroll work, a marble dresser top with a mirror, and a six-piece Winthrop silver coffee and tea service from Reed and Barton of Taunton.

Family members often passed along Elizabeth's detailed reports of trips to New York, Philadelphia, and Baltimore as well as their attendance at concerts and the opera. When at home, the couple entertained the family and a wide circle of friends with club gatherings, whist, charades, parties, dances, and musical presentations. Their private balls, which required more space, were held at Washington or Pantheon Hall on Main Street. Elizabeth's sisters never missed a chance to be included at the most elegant socials of every season, and occasionally sister Mary was a featured soloist. The senior Pinkhams were always included, and Seth was polite but firm in insisting that his girls be known as Captain Pinkham's daughters, not Elizabeth's sisters.

During the age of wealth and privilege that resulted from the whaling industry, another extravagance became common among the well-to-do: a summer house at 'Sconset, on the far eastern reaches of the island, where families could "forget the focus of fashionable life [and celebrate] the fine bracing air, the excellent water, and the unique customs of the place—all admirably adapted to refresh and invigorate both mind and body."

In the early 1800s, Captain William Brown bought a structure in town that had been a sailcloth factory, moved it to Madaket for a boathouse, and still later, with further modifications, moved it again in carefully marked pieces twenty-five miles to 'Sconset for a family summer retreat. Written accounts agree with long repeated family lore that the post-and-beam house boasted a first life of old-growth timber, and a second life thanks to beams and planks salvaged from the British ship, *Queen*. Over the years, a broken mast, sets of spars, a gang-plank, bricks, windows, and wooden debris from other wrecked ships were added to the house. Located

The house at One Pleasant Street, 1890s. *The Greek Revival was built by William and Elizabeth Crosby in 1837.* –Nantucket Historical Association

"Sans Souci"[5] at 26 Broadway in 'Sconset, 1887. *Outside the summer home of Captain William Brown are, left to right, Elizabeth Plaskett Bennett holding Florence Bennett (Anderson), Mary Pinkham Plaskett, Mary Brown, wife of William F. Brown, and Elizabeth Pinkham Crosby.* –Nantucket Historical Association

at 26 Broad Street and featuring thirteen different windows, the house became the eclectic summer home of the Browns, the Pinkhams, and the Marshalls. They called it *Sans Souci*. It had a great room on the first floor with two adjacent bedrooms, two bedrooms on the second floor at the top of a steep stair, and a wart, or attached shed, that served as a kitchen. A very short walk down the steep lane to the beach and not many steps from the town pump, *Sans Souci* was the locus of all the children's happiest memories. Malvina remembered Grandfather Brown sitting in the house and handing out apples picked from his orchard to the young ones; now she and Joseph were eager to share it all with Nelly, a member of the fourth generation at *Sans Souci*.[5]

They were bound by family and the commonality of family places, but on Sundays they were part of a larger gathering—a congregation. The graceful, Quaker-gray Unitarian Church[6] occupied a slight rise on Orange Street that looked out over the town. Instead of the more traditional New England tapered steeple, it had a large, square tower (later a four-faced clock tower), where the fire watchmen could stand and signal the townspeople with a lantern in the direction of a blaze. Above that structure was a belfry that held the pride of the island, the Portuguese bell purchased by Captain Charles Clasby in Lisbon in 1812, which had been delivered to Nantucket by Captain Thomas Cary. Regardless of religious affiliation, the people of Nantucket "lived" by the morning, noon, and evening chimes, the island's "sweetest sounds," which came from a bell originally planned for a Catholic convent. Finally, the "crowning glory of the church exterior was a gold dome that shone in the moonlight as well as the sunshine." Inside, the "gray and silver of the walls were outlined in the pulpit alcove with *trompe l'oeil* painted by Carl Wendte in 1844 giving the sanctuary a look of much greater length, and the long gray shuttered windows contrasted with the mahogany pew rails and ivory white woodwork."

Although Seth Pinkham's pew was "in the south transept, third from the pulpit" when the side galleries were removed during alterations in 1844, the Pinkham pew remained in the third row but now faced the pulpit.

After being away so long, Malvina was particularly comforted to see that Rev. Edes was still officiating, since she and Joseph had received word that he was planning to transfer elsewhere. Nelly liked sitting in grandfather's pew, even though she had never known her grandfather. Still, the service was rather long for a wiggly five-year-old, even if she did find the bright-red bench cushions comfortable and enjoyed turning around frequently to smile at Auntie Harriet and Auntie Helen, who were seated in the Pinkham's second pew. So, Nelly was often allowed to stay at No. 46 with Mrs. Pompey while the others attended service. She would help prepare the Sunday repast by greasing the pans, licking the mixing bowls, and asking questions about everything.

On First Day each week, the Marshalls learned of old friends and extended family members who had moved off island and exchanged news with those who had loved ones at sea in the far reaches of the Pacific—places whose unusual names were so familiar to Nantucketers—and learned about the growing number of personal hardships that resulted from the continuing decline of the whaling industry.

Malvina and Joseph were distressed by the significant recession that had occurred in Nantucket since their marriage in 1845. At that time, Nantucket merchants had eighty-six ships at sea as well as several brigs and schooners: Nearly half of the whaleships in the world sailed from the island, and the docks and harbor saw a steady stream of ships delivering cargo or preparing to leave for the Pacific. The wharfs, with their ropewalks, warehouses, cooperages, and forges were crowded with workers while wagons and handcarts rumbled from one wharf to another.

As many as thirty-five candle factories had been in operation back then, and a goodly number of sail-makers, brickmakers, and foundries further helped to fuel the island economy. Stores were full of exotic goods: spices from India, nuts and fruits from South America, inlaid furniture and fine table settings from China, French wines, merino shawls, Azorean oranges, and Spanish olives. Construction—especially the development of stately mansions for sea captains and merchants—was at an all-time high.

But in 1846, the Great Fire gutted the town, destroying many homes and commercial properties, and driving many businesses into bankruptcy or unsustainable decline. William Crosby's oil and candle company lost its factories, warehouses, inventory, and storage facilities, while the cooperage owned by Obed Marshall, located on the ground floor inside one of the sail lofts, was destroyed.

And despite remarkable attempts at recovery, the Gold Rush of '49 emptied the town of almost half its men of working age, and forced many ship owners to convert vessels to carry passengers and cargo to Panama and California. Several new business ventures like silk manufacturing and shoemaking attempted to revive the economy but were short-lived. And although the island had long been waging a valiant battle against the rising sand bar at the mouth of the harbor that kept newer and larger ships from entering—first by dredging and then by floatation devices called camels—it soon became clear that most cargoes would have to be unloaded at mainland ports such as New Bedford. Finally, the discovery of "earth oil" (petroleum) and processes to refine it into kerosene for lamps proved a death sentence for the more expensive whale oil. Yet the whaling industry continued.

During 1856, in the course of their daily errands and walks with Nelly, Joseph and Malvina saw the harbor almost empty of seafaring ships and the wharfs gravely lacking in workers and equipment. Main Street was quiet; some shops were closed. Only the new planting of elm trees—a gift from the Coffin brothers—and the new street lamps indicated a guarded optimism for the future. The usual masses of seamen from all over the globe—Samoans, Portuguese, Malays, Kanakas, Cape Verdeans, and Azoreans—were no longer seen on the docks or on Main Street; they had left to seek work in New Bedford and other ports that were still active in the whaling trade.

But the country roads and sleepy outlying villages of Nantucket had changed little. The Marshalls borrowed William's horse and carriage to visit again in Madaket, Polpis, and Wauwinet, and they had family picnics on the grasslands at Squam. They watched with pleasure as Nelly ran on the beach from one pile of dried seaweed to another, looking for shells and pretty stones that might be attached. Nelly was also thrilled to visit the dairy, watch the cows being milked, and play with the kittens in the barn. Sometimes Joseph would hire a saddle horse—always a spirited black one—and ride at full gallop on the straight stretch of Milestone Road; and on occasion, he would take Nelly in the saddle in front of him for short rides closer to town. In June, the family went out to Miacomet Pond for the annual sheep-shearing and watched as the sheep were rounded up for their once-a-year haircut. That July, Joseph traveled to Westport to meet with his owning partners in the new ship, *Aurora*, and contract with his officers, while Malvina, Harriet, Helen, and Nelly went to Boston to shop. When they all returned to the island, No. 46 became the center of preparations.

The Marshalls had been home for more than a year, but it would be only a few months before their scheduled departure in November. They carefully made lists of things that they would need to take, and Malvina agonized over the furniture for the stateroom and the large captain's cabin—furniture that would be delivered directly to the ship. She purchased new linens, blankets, and quilts; added new pieces to her own tea set; and collected all the books that she and Joseph would want with them. She packed personal things for herself, and for Nelly, she packed books, extra paper and pencils, a globe, needles, embroidery linen, and thread—all the necessities for a little girl's formal schooling. With only two months left on Nantucket, Malvina set aside extra time to spend with her mother and visited Joseph's mother often while he was engaged in meetings with William, who was to continue as business manager for the *Aurora* voyage as well as executor of the Marshalls' estate.

The days flew by, and soon it was time for Joseph to leave for New Bedford in order to recruit the crew and supervise the loading of the ship. Malvina and Nelly left two weeks later with tearful goodbyes. And so it was that the Marshalls left their island home—thankful for their ties to family, friends, and church, and hopeful for a successful voyage and a safe return to this place where, as Malvina described it:

> *There is a unique fragrance in sea air transported over the moorland sweeps with whiffs of bayberry and sassafras, sweet fern, arbutus and violets in their season, dwarf pine and scrub oak, huckleberry and beach plum bushes, wild lilies, marshmallows, orchids hidden in aromatic swamps, and various sturdy heath blossoms.*[7]

Helen "Nelly" Marshall's shipboard toys. *Among the items in the collection of Nantucket Historical Associations are: Nelly's Izannah Walker doll, her blue jump rope with ivory handles and turk's heads, "The Fox and Geese" book, and Malvina's measuring tape.* – Photograph by R. S. Brown; objects from Nantucket Historical Association

Chapter Seven

Aurora

She is daybreak and comes from the East representing the peoples from Eastern lands. Every morning Aurora rises from the sea and rides in her horse-drawn chariot on large white wings across the sky to the West ahead of the sun to bring forth a new day. She is sister to Luna the Moon Goddess, her brother is Sol, the Sun, and her children are the four winds. Such is the power of Aurora.[1]

On November 10, 1856, the *Aurora* of Westport—a ship built of carefully selected hardwoods, propelled by new white canvas sails, and carrying thirty crewmen, four mates, the captain, and his wife and child—left New Bedford for the Pacific Ocean. Its quest was whales. The pilot was discharged at Gay Head Light, and on November 12th, Joseph ordered all sails set and charted a course north into the Gulf Stream, east toward Europe and then south to the Cape Verde Islands off the coast of Africa. From the outset, it was obvious to both captain and officers that the *Aurora* was an exceptional bark: she was fast for a whaler and responded with precision to every quarterdeck order and expectation. The improved design of the davits facilitated a safer and more rapid lowering of the boats and the installation of a crow's nest[2] with railing, vertical staves, and leather hatch flap provided enhanced protection for those standing watch. A large galley fitted with the latest in stove and oven equipment proved its worth when, in the third week of the month, a New England Thanksgiving dinner including corn bread, turkey, and pumpkin pudding was served to all, following the prayer service on deck. And although no whales had yet been taken, the holiday mood continued when, only days later, tiny pieces of Nelly's birthday cake were passed around.

Now five years old, Nelly was excited to receive a red-seated rope swing from her father, who dutifully hung it from a temporary line installed between the main and mizzen masts, allowing for a "thwartship's arc" within the safety of the quarterdeck. Another package wrapped in brown paper awaited Nelly's inspection. For weeks, Malvina had been secretly working at night on the birthday surprise for her daughter, who wanted to dress like a real "sailor." Scavenged from the ship's slop chest were enough navy-blue wool, padded linings, and black buttons with anchors to create a perfect child-sized pea jacket[3] with a large wind-breaking collar. Malvina looked forward to the unwrapping of the little coat, which would also be a practical necessity, as the *Aurora* would soon be sailing into the cold, windy climate of Cape Horn. But to her considerable dismay, Nelly responded to the gift with a polite yet cursory "thank you" and went back on deck to play on her swing.

It was not until several days later and many prodding questions that Nelly found the courage to tell her mother that the pea jacket did not look like the sailors' coats. It did not have patches, she lamented. Malvina had plenty of time to make corrections, and later, when the first chilly wind blew across the play area on the quarterdeck, the coat was replete with yellow gingham squares on each elbow.

Fine weather in the South Atlantic allowed for frequent lowering of the boats to train the crew, and by December, when the *Aurora* sailed into the wide natural bay at the Cape Verde island of Sao Vicente, Joseph was pleased with the progress of the boat drills. He left the ship and went ashore to hire two recruits and replenish food and water supplies. When he returned, the deck had been cleared for laundry[4] to be done in the warm, dry air, and the crew was given free time to go swimming. Yet precious days were passing, and still no whales had been taken. A course was set for Pernambuco, Brazil, and just before Christmas, the *Aurora* had a gam[5] with Capain Jeremiah Ludlow of the bark *William Hamilton* of New York; the Marshalls served trays of cheeses, yeast breads, and sliced ham with cider sauce and lemonade.[6]

Ludlow had experienced a long and difficult passage around the Horn, during which he lost a boat, had several sails badly damaged, and had taken on water through the leaking forward hatch. But he and his crew had caught and tried out their last whales off the northern coast of Brazil and were now on the way home. When the captain offered to carry mail to

Excerpt from the "Panorama of Recife,"[7] by Frederick Hagedorn, 1855. *Whalers referred to Recife as Pernambuco, the state in which Recife is the capital. After whaling in the fertile bays off the Brazilian coast, whalers often made port here—sometimes for respite, and often to trade. Ships sold oil in Pernambuco and loaded up with commodities such as coffee and sugar. Located at the eastern extremity of Brazil, Pernambuco was a popular place for mariners to get provisions and do maintenance in preparation for the long journey home—normally between 90 to 140 days.*

New York, all on board the *Aurora* who could write hastened to prepare letters. Malvina and third mate George Mills helped some of the younger sailors with spelling and writing addresses on the envelopes. And after Captain Ludlow's report of recent storms, Joseph was further convinced of the wisdom of going directly to Pernambuco, where the *Aurora* could be prepared for the always-challenging voyage around the Horn to Valparaiso.

Although eager and hard-working, the crew represented a diverse racial and ethnic mix; and since at least half of the men did not speak English as a first language, misunderstandings and prejudices sometimes led to problems. On December 23, after the mates had broken up several fights, "the captain went forward [to the foc's'le] and delivered to the crew a cantriver[8] in respect to their lives." He explained the necessity of working together not only in the boats but also on the ship, making it clear that in the days ahead their lives and fortunes depended on one another. He further stated his intention of instituting the discipline necessary to achieve strict order. Then he repeated his speech of caution in Spanish. The deportment of the crew soon improved, at least for the present, and Joseph was able to focus his efforts on improving the navigational and leadership skills of his officers. He reviewed the uses and sequences of the sextant, chronometer, compass, charts, maps, and calculations. He made certain that each mate stood a watch, during which Joseph took a lunar observation as well as regular solar observations, and he enlisted Malvina's help in giving the mates some rudimentary instruction in Portuguese vocabulary and phrases.

A careful inspection of the ship was undertaken and painting began. On December 28, three blackfish were caught and tried out, while routine jobs all over the ship kept every sailor busy. By the first week in January, the ship was cleaned of the grime left behind from the greasy, black smoke of the try pots, the barrels were stowed below, and on January 16, the early watch landed two skipjacks that added eighty pounds of tuna to the week's evening meals. The *Aurora* was a day away from the major port of Pernambuco when the captain posted a prize of a shirt, tobacco, and cash for the next whale taken on the way to the Pacific—a strong motive not only to encourage initiative, but more important, to insure the crew's return to the ship after liberty ashore.

Those on the *Aurora* who had not been to Brazil before were stunned by the hundreds of sailing craft in port from all over the world, and on land they were

A colonial fishing settlement established in 1537, Recife is encircled by a coral reef that protects ships inside the harbor. The United States Consulate was established in Recife in 1815; making it America's oldest diplomatic post in the Southern Hemisphere. Artist Frederick Hagedorn's panorama was created in three circular panels, each measuring about 9.25 feet in length by 2.25 feet wide.
– From a lithograph of the original watercolor, Wikimedia Commons

amazed to see the crowds of people everywhere. The first stop for Joseph was Brazilian Customs, after which the Marshalls went together into the United States Consul's Office. While Joseph was presenting the required documents and filling out what seemed like a never-ending pile of official forms for the customs agent, his wife and daughter were excited when mail for the *Aurora* was delivered to them. Letters from home had miraculously arrived, addressed to the intended recipient, Bark *Aurora,* Atlantic Ocean. Sometimes the envelope also specified "Outward Bound" or "US Consul." There were boxes of food from Boston for Captain Joseph Marshall, Bark *Aurora,* Pernambuco, and there were letters and boxes from Nantucket for Joseph and Malvina containing newspapers, magazines, books, and bolts of cloth.

Of course, there was mail for others on board the *Aurora* as well, but one box was different—it was from Providence, Rhode Island, and it was addressed to Nelly Marshall, Bark *Aurora,* Atlantic Ocean. This was Nelly's first personal mail, and she could read her name on the box. Although the crowded offices of the consul afforded little space to open gifts, Nelly was allowed to sit on top of the shipping cartons, and after Joseph broke open the water-tight carton, Nelly was able to peel off the layers of wrapping paper and take the cover off the pink and white striped box. Inside, under carefully folded linen, was the pretty face of an Izannah Walker doll.[9] She had a fair complexion, shiny cheeks, and painted face framed by dark brown painted curls. Her plain full-length dress of tiny pink flowers printed on ivory cotton covered her white cotton underclothes, and her brown painted boots could just be seen beneath the hem. Nelly had never seen such a beautiful doll, and it was larger than any she had ever had.

Inside the cover of the box was a note explaining that the doll was a gift from Auntie Elizabeth and Uncle William and was purchased in Providence by Auntie Helen. Nelly hugged her new doll and announced that she would call her Nelly. Back on board, while Joseph, first mate John Berry, second mate George Hussey, and third mate George Mills were employed in overseeing every detail of the ship, Malvina and Nelly had their own work to do. Nelly had begun her lessons during the family's Nantucket sojourn and already could print her name, read all the numbers on her blocks, and count to twelve using her mother's measuring tape. But now the time had come for more grown-up material, and so every morning between ten and noon on the quarterdeck in good weather, or in the captain's cabin on rainy or cold days—Malvina and Nelly "went to school" with a curriculum of reading, recitation, arithmetic, music, needlework, and history. Nelly liked history best because it was a series of stories about people she did not know and she very much liked listening to stories. Her favorite was from the Bible—*Moses in the Bulrushes.* She found arithmetic curious when Malvina showed her that "an orange cut into two pieces was greater than an orange cut into four pieces."[10]

Nelly's first assigned recitation was chosen to please her father; it was about a blacksmith who worked long hours at his forge: "Under a spreading chestnut tree the village smithy stands…" And when Joseph and the mates listened to the diminutive five-year-old with an unusually low pitched and powerful voice, and standing on a shipping crate, they applauded her performance. Though proud of her accomplishment, Nelly thought it was hard work to practice and get all the words right. But she did understand that school was necessary and that her cousins on Nantucket all went to school. Still, the sailors often captured Nelly's attention as they climbed the rigging, rolled and stowed the barrels, and washed the decks. When it was time to practice her needlework, she pretended she was mending sail, and when she worked on sums and take-a-ways, she loved to use the colored beans that the steward had given her.

Malvina, besides being a mother and teacher, had work to do. She made Joseph's shirts and had recently begun letting out the seams in his outer clothes and carefully moving buttons to accommodate the weight he had gained as a result of giving up tobacco. She was also the family barber and, drawing from articles in magazines from home, was able to keep up with both men's and women's styles. Nelly's haircuts were more than a matter of styling, however. Concerned about their daughter's size and health from the very beginning, Joseph and Malvina subscribed to the practice of regular haircuts to promote physical well-being.

Accordingly, it was no surprise that the responsibility for simple medical care of the officers and crew—such matters as cuts, sunburns, insect bites, rashes, and fevers—was assumed by Malvina. According to records, "she was also a self-appointed hygiene officer for the *Aurora*," and undertook regular removal of roaches and sanitizing the ship's living areas with bleach and borax. Coming from a sea-faring family, she was familiar with illnesses, like scurvy, that were common to nautical life, and when possible, sought to modify the shipboard diet to improve health. Of course, the more serious medical problems were the responsibility of Joseph, who had become adept at suturing wounds, pulling teeth, and even setting bones after a fall. Like his wife, he was a great believer in preventative care; he even had several jump ropes made for the crew and encouraged them to exercise regularly.

By February 12th, the *Aurora* was well on its way to Cape Horn, following the coastal route west of West Falkland Island and past New Island, where birds and intermittent squalls of wind and rain surrounded the ship. The skies cleared as they approached the tip of South America, and Joseph was able to take navigational sightings off Staten Island (Isla de los Estados). But almost immediately, heavy seas arose and threatened the progress of the *Aurora* as well as the *Hector* of New Bedford, the *Cicily* of Thornburg, and a third whaler from New York—all headed for the Pacific. On February 17, George Mills wrote in his journal:

> *The storm increased with the night and the sea was lashing into tremendous confusion. There was a fearful sullen sound of rushing waves and broken surges. At times the black volumes of clouds overhead with vivid flashes of lightening were doubly terrible. The thunder bellowed over the wild waste of water and I saw the ship shaygaring and plunging among the roaring caverns. It seemed miraculous that she regained her balance or preserved her buoyancy. When she would dip her bow almost buried beneath the waves. Sometimes an impending surge appeared ready to overwhelm her. And nothing but a dexterous movement of the helm preserved her from the whistle of the wind through the rigging, the groaning of the bulkheads, creaking of the mast. The ship wallowed in the sea and was frightful.*[11]

For almost a month, Joseph and the officers and crew guided the *Aurora* through the Cape storms of foul and frigid weather with only a single mishap: James Harper, a foremast hand, fell from the yard but was not seriously injured. Finally, they left the notorious weather of the Cape behind, and one day in the first week of March, the deck officer, first mate John Berry, was overjoyed to sight the familiar outline of the Juan Fernandez Islands, west of Valparaiso. Joseph's first item of business was to send a boat ashore for eighty barrels of potatoes—then he broke out molasses, bread, and salted-meat for the crew while Malvina poured grog. All were thankful for the safe trip west and the dawning of a new day. The Pacific lay before them, and there were whales in that blue-green water. The hunt would soon begin.

Crew list of the bark *Aurora* of Westport, 1856, Joseph Marshall, master. *The mates were John Berry, George Hussey, George Mills, and Daniel Nevers.* – Nantucket Historical Association

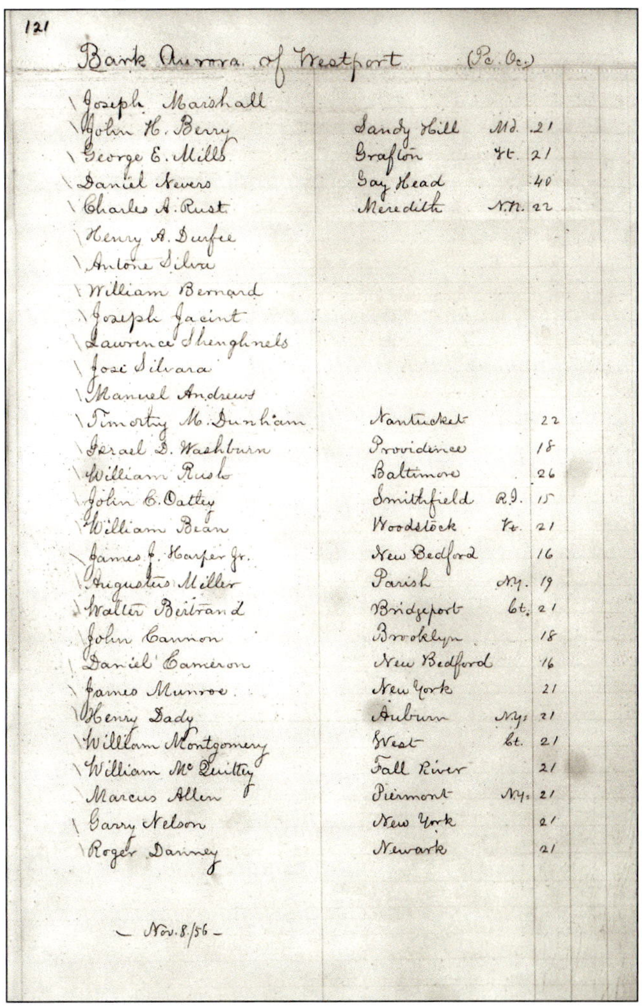

121

Bark Aurora of Westport (Pe. Oc.)

Name	Residence	State	Age
Joseph Marshall			
John H. Berry	Sandy Hill	Md.	21
George E. Mills	Grafton	Vt.	21
Daniel Nevers	Gay Head		40
Charles A. Rust	Meredith	N.H.	22
Henry A. Durfee			
Antone Silva			
William Bernard			
Joseph Jacint			
Lawrence Shenghnels			
Jose Silvara			
Manuel Andrews			
Timothy M. Dunham	Nantucket		22
Israel D. Washburn	Providence		18
William Rush	Baltimore		26
John C. Oatley	Smithfield	R.I.	15
William Bean	Woodstock	Vt.	21
James J. Harper Jr.	New Bedford		16
Augustus Miller	Parish	N.Y.	19
Walter Bertrand	Bridgeport	Ct.	21
John Cannon	Brooklyn		18
Daniel Cameron	New Bedford		16
James Munroe	New York		21
Henry Dady	Auburn	N.Y.	21
William Montgomery	West	Ct.	21
William McQuilty	Fall River		21
Marcus Allen	Piermont	N.Y.	21
Garry Nelson	New York		21
Roger Dainey	Newark		21

– Nov. 8/56 –

Captain Joseph Marshall letters. *Mail was addressed to ships and sometimes to consuls in cities. Some were simply sent as "Captain Joseph Marshall, Bark Aurora, Pacific Ocean." Nelly Marshall dropped a letter to her brother in the box on Charles Island (Galapagos) addressed as "Joseph Marshall, Nantucket," and it arrived safely.* –Nantucket Historical Association

Chapter Eight

Profiles from the Pacific: A Plantation Owner, a Prince, and a Queen

A prevailing mood of thankfulness for safe delivery from the challenges of Cape Horn escalated to one of mirth and merriment on board the *Aurora*. As soon as the delivery of potatoes arrived, sails were raised for a swift 400-mile trip to Valparaiso, and after dinner, several sailors brought a fiddle and accordions to the forecastle deck. When the Marshalls joined them, they played a spirited medley of familiar tunes, beginning with *Darling Nelly Gray* and *Pop Goes the Weasel* and continuing on to *Yankee Doodle, Oh Susannah,* and *Camptown Races.* To everyone's surprise, Joseph led Malvina into the center of the merry crowd and danced a quick-step followed by a polka to the even faster tune of *Jenny Lind.* The sailors began to dance with impromptu shuffle and tap steps, and even Nelly was spun in place by second mate George Hussey. Finally, a sailor from Maine captured everyone's attention with his energetic performance of the sailors' hornpipe, topped off with the traditional spit over the shoulder.[1] At daybreak came the thrilling call from above by 15-year-old John Oatley,[2] "Thar she blo-o-ows!"

The boats were quickly and quietly lowered, and the crews gave chase. It was a sperm whale—a big one. But they faced a dilemma: while they needed and wanted that kill, they were also very low on food and water, and the delay to process the whale meant it would be days before they arrived in port. Fate became the decision maker, however, for the whale was taken with little effort and brought alongside for the cutting in. Everything was done to speed the process: both officers and crew bore a hand at several tasks and worked, as fast as possible, while the cook set out food to be ready when a break in the men's work allowed them to eat. Realizing that the whale would bring more than 100 barrels, Joseph promised cash rewards for the accelerated work. Malvina was enlisted to assist the steward in the galley, and Joseph assigned the handling of the *Aurora* to the third mate and the shipkeepers—the cooper, the blacksmith, and the cook—while he supervised the rendering of the whale.[3] As the try pots spewed out their greasy black smoke, it began to rain, and enough water was collected to ease the threatened shortage. The only mishap in the ensuing days of difficult and dangerous work occurred when the third mate accidently broke the crystal in the captain's watch.

When the *Aurora* finally entered the port of Valparaiso, the decks were still unkempt and barrels were still being stowed. But the port facilities in Valparaiso allowed easy access to the city, and as soon as the ship was thoroughly cleaned, the officers and crew could enjoy a scheduled liberty. The only grumbling arose from the Captain's decision to withhold the promised bonus money until the *Aurora* had left port.

Valparaiso in 1857 was the busiest seaport in Chile. It was the first major stop for ships coming from Europe via the Cape of Good Hope, and from Europe and North America via Cape Horn. As a result, it was more European than the smaller Talcahuano to the south; as a center for English, German, and Italian shipping, it allowed for business to be carried out in English—a great advantage for American and British whalers and sea merchants. But Valparaiso was an old city founded in the early 1500s and was one that had suffered from military attack, fire, and earthquakes. Because thousands of people, horses, carts, and domestic animals thronged the streets, the resulting debris and animal waste produced an unpleasant environment.[4]

George Hussey escorted Malvina and Nelly to town and made lodging and dining arrangements for the Marshalls. Despite the condition of the streets, Malvina was glad for the chance to walk, visit museums and the cathedral, and purchase personal supplies for the family. Nelly liked ordering food in the hotel, and the great expanse of the city plaza prompted her to learn to skip. The *Aurora* would be in port for at least a week

Whaleships navigate the treacherous floes of Cape Horn at the tip of southern Chile, 1845. *In 1791–1793, the* Beaver *of Nantucket and the* Rebecca *of New Bedford became the first American whaleships to double Cape Horn, returning with full loads of oil.* – From Russell and Purrington's Panorama, New Bedford Whaling Museum

for repairs, caulking, painting, and cargo loading, and during the stay, three sailors were recruited and one was dismissed. All the while, Joseph was anxious to get to sea.

After leaving Valparaiso, the *Aurora* first sailed southward two hundred miles off the coast heading for the Archer whale grounds, then set a course north and west around the Juan Fernandez Islands before aiming toward the Galapagos Islands, off the coast of Ecuador. But weeks turned into months with no sight of whales. Two gams—both with other whalers out of New England—broke the monotony of the seemingly empty sea. The first was with the ship *Courier* of Nantucket; Malvina's old schoolmate, now Mrs. Coffin, welcomed her on board for that visit. The second, however, nearly ended in tragedy. Joseph, Malvina, and Nelly had visited with the Swains of the whaler *John A. Parker* and had spent a fine afternoon exchanging mail, books, newspapers, and recent news.[5]

But when the family left the ship, they encountered a frightening and dangerous situation:

> *They said their good byes and their boat headed off toward the Aurora, darkness fell soon [quickly] as it was wont to do in those latitudes and suddenly the captain and the men realized there was no light on their ship and that the ship was not even in sight. There was no moon but the stars were bright as always in the southern seas. [The men] continued to row in the direction where the ship was last seen. The Parker had gone on her way never dreaming there would be any trouble. Everyone in the boat was very still save for Joseph who gave occasional directions to the men. As they [finally] went up the side of the ship and stepped onto the deck, Joseph sent his family below. The mate in charge tried to give an excuse for not setting a light but was immediately discharged and [later] transported to the nearest port of Callao, Peru.*

When the sailors had free time, especially when the ship was becalmed, they engaged in a variety of activities. Scrimshaw or scrimshont work was a favorite pastime. Some would carve whales' teeth into cane handles, umbrella handles, a dog's head, a horse's head, a simple curve, or a knot. They made circular jagging knives for edging pie crusts or etched pictures of full-rigged ships, shields of the United States, the American flag, or a ship's female figurehead on a whale's tooth.

Sailor's palm. *Example of a leather palm used by sailors to protect their hands and to push the needle into the canvas when mending sail. A miniature palm was made for Helen Marshall by a sailor on the Aurora.* – Photograph by Jann Gentry

The only implements used [to prepare the tooth] were a jackknife, a fine saw, and sandpaper. For etching, they used a needle and then rubbed in India ink or sealing wax. One of the most beautiful gifts made by a sailor and presented to Malvina was a tatting shuttle crafted from ivory and silver ten-cent pieces, and surrounded by a slender thread of dark fiber from a coconut shell. It was inlaid with silver hearts on one side and diamonds on the other. The sailors also used their needles and ink to decorate each other with often elaborately designed tattoos. Nelly was fascinated with the procedure and took careful note of how it was done. One day when the sailors had taken a break from their crafts and left a needle and ink out, Nelly began to adorn her wrist with a permanent bracelet and managed to prick several "inked beads" before Malvina caught her.[6]

Another way the sailors passed many long hours was called "sailor's comfort," where they took everything out of their sea-chest, looked it over, and put it back. Typically, there was the little box of spools, needles, darning thread, a piece of wax, a pair of scissors, a thimble, a little bag of buttons of all sorts, and one or two pieces of calico to mend a shirt. Often a daguerreotype or two would be carefully wrapped in paper along with two or three letters from home—and, of course, there would be paper, pencil, and a pen to answer the letters from some distant port of call.

Of course, part of the sailors' work was to mend sail and to whip and splice rope. They would use a specially designed "palm" to help push the needle and heavy thread through the canvas or rope. Made of leather and fitted around the hand, it covered the palm and had a hole for the thumb. A small disc of metal with shallow dimples to keep the needle from slipping was fitted within a larger circle of heavy, molded boot leather and sewn to the palm slightly to the right of center. A sailor's palm was a prized possession, painstakingly crafted to fit the hand that used it. After many requests, Nelly was finally outfitted with a working sailor's palm of her own and spent many happy hours sewing with her smaller needles and light fabric.

It was October when the *Aurora* arrived at Tumbez, on the north coast of Peru. It was a very different port and "had to be entered by boat through the breakers that rolled over the shallow place at the mouth of the river." After months at sea, the sweet smell of land, along with the sight of tropical trees and lush growth, provided a special welcome for all on board. The Marshalls went first to the marketplace, where they purchased luscious fruits and vegetables, some of which "were strange to look at." And the scents of the marketplace "…were mingled with tobacco—an odor not found elsewhere save Cuba."

When the ship's stores had been purchased, Malvina returned to the *Aurora,* while Joseph left on horseback for a trip up the river to Somontez Plantation with Nelly in the saddle in front of him. It was a long ride and the road along the river was dusty, but the air was fragrant with the wild blooms of the jungle. The plantation house was typical of the country: it was perched on logs and thatched with banana leaves, with sides covered by matting that could be raised or lowered and partitions made of bamboo.

Señor Somontez and his wife and daughters were most hospitable. Although this was a country house, it had sterling settings on the table. Señora Somontez and her adult daughters were dressed in elegant formal attire. They wore silk dresses of bright colors enhanced by voluminous layers of petticoats and adorned their hair with tall, elaborate combs that provided a framework for black lace mantillas, which draped down over the shoulders and cascaded all the way to the floor. They also wore jeweled crucifixes and a wide assortment of bracelets and rings.

Servants, whose duties included holding up the trains of the dresses when the ladies went walking outside, did all the work.

After a fine midday meal was served, Joseph and his host retired to discuss customs requirements and costs. As Joseph would later write, "Somontez was a courtly Spaniard whose power over the government agents every sea captain was glad to use and pleased to pay for." Somontez explained his fee to Joseph in somewhat couched terms: "Oh Captain, just put an ounce over each eye and there will be no trouble."

These ounces in Spanish currency were the equivalent of $32 and would be a small price to pay for avoiding a trip to Payta. Moreover, Joseph and Somontez had become more than acquaintances or business colleagues over the years. They enjoyed each other's company and their visits were always cordial.

> *[Somontez'] abundant hair was silver, and he was dressed in the latest southern European fashion: He wore high waisted, pleated silk britches with black embroidery around the side pockets, a red silk shirt with vertical stitching and a wide cravat tied loosely in a bow. Over the unmatched garments he wore a satin coat with contrasting collar and cuff trim, which fell to calf length.*[7]

Everyone in the Somontez family was appreciative that Joseph included them in social discussion by speaking in Spanish as much as he was able. Joseph left the plantation in the late afternoon, but it was not long before Nelly fell sound asleep, and when they reached Tumbez, she had to be carried from the saddle to the ship's boat for their return to the *Aurora.*

The next day, with the demi-johns[8] filled and the supplies loaded, the course was set for the Galapagos Islands to the northwest en route to Hawaii. They must find whales, and a cruise among the archipelago of Ecuador seemed promising, as it had always been successful for American whalers since the Nantucket ship *Beaver* first rounded the Cape in 1791, and since the *Globe* discovered the off-shore whaling grounds in 1818. But, although the winds allowed for fast travel, the weather became hot, the crew became disillusioned by the scarcity of whales, and the mates began to worry about the need for a longer-than-usual voyage. Even the newest green hands could quickly calculate the ratio of barrels stored to time at sea. As the *Aurora* neared Charles Island on the south-central border of the island group, the skies became overcast, the temperature dropped, and a light rain began to fall.

The Galapagos chain was comprised of eighteen major islands and would become well known for providing Charles Darwin with the biological and geological data that would eventually lead to his theories of natural selection. It also was situated in the Pacific breeding grounds of sperm whales and provided whaling ships with opportunities to obtain fruits and vegetables, hunt for land tortoises or terrapin, and take on water.

Finally, the *Aurora's* luck turned, and her boats were sent in pursuit of three whales. Two were killed and securely chained alongside for cutting-in, but one was lost while being towed back to the ship. By the third week in November the work was done, seventy-five barrels were stowed, and the cleaning-up process was well underway. Although the whales were small, they were sperms and therefore added to the ship's valuable cargo of bone and teeth as well as quality oil.

A week later, Nelly celebrated her sixth birthday, and her parents promised that if she finished writing a letter to her brother, they would take her to the post office and let her mail it herself. So Joseph sailed into Post Office Bay on the northwest coast of Charles Island, dropped anchor, and lowered a boat with Mr. Hussey, four oarsmen, and a very excited Nelly. When they went ashore, they headed directly to the box mounted on a post, known and used by whalers for

Gigantic land tortoises from the Galapagos Islands. *The giant tortoises of the Galapagos Islands were a vital food source for sailors—often referred to as terrapin or turpin by the crew who hunted them.* – Etching by Evremond de Berard in 1864 for Charles Darwin's *A Naturalist's Voyage Round the World*, Library of Congress

Charles Island (Floreana Island) of the Galapagos Islands, 1835. *Fresh water, food, and a post office—consisting of a mail box on the beach—made an island visit essential. The mail box is still used to post letters today.* – Etching by Evremond de Berard in 1864 for Charles Darwin's *A Naturalist's Voyage Round the World*, Library of Congress

over fifty years.[9] Mr. Hussey sorted through the letters and took the ones that were addressed to destinations to the west—especially to Hawaii, and then held the lid open for Nelly. She looked proudly at the envelope she had addressed all by herself: "Joseph W. Marshall Nantucket." And then she dropped it in.

Several storms, followed by lack of winds, impeded the trip to Hawaii. So instead of the planned Christmas arrival, it was mid-January when the *Aurora* put in at Honolulu for repairs. At the end of that month, the ship finally entered the roadstead of Lahaina, Maui, and dropped anchor among "the forest of masts from around the world." This popular port had been the capital of Hawaii until 1852, and because it was located on a sheltered channel, there was always enough wind to enter and leave without the need of a pilot.

Lahaina was lush, green, and fertile, with an abundance of water, crops, trees, and grass. American missionaries and medical missionaries lived in houses that reminded the homesick crews of their life back home. There were also comfortable-looking, grass-roofed adobe houses, steepled churches, stores, a customs station, a lighthouse, and a jail that was built specifically for rowdy sailors who did not adhere to the curfew laws requiring them to return to their ships at night.

Joseph and Malvina had planned a layover at Lahaina before going on to the farthest reaches of their voyage, which included the Marquesas and the Society Islands of French Polynesia. They rented a house within walking distance of the water in an area where many American families were living. It had large windows, a traditional grass roof, and three rooms, including two bedrooms and a very long, rectangular sitting room with a door on each end. It belonged to a British merchant named Marchand and was nicely furnished. Meals were served to the Marshalls by servants who also took care of the house, the laundry, and a small garden.[10]

While in Lahaina, Malvina spent many hours visiting with friends and acquaintances from Nantucket, New Bedford, and earlier gams on the *Sea Queen*. Joseph, meanwhile, undertook a challenging task: teaching Nelly to swim. They walked to an area where the beach was flat and the water was shallow and without active waves. Nelly wore a short-sleeved play suit with bloomers made for the occasion, and Joseph, although a competent swimmer, observed

Port of Lahaina, Maui, 1854. *Lahaina was the capital of the Kingdom of Hawaii from 1820 to 1845 and was the hub of the Pacific whaling industry. New England whaling families often laid-over on the island, where food and fresh water were abundant.* – Illustration by Edward T. Perkins, Wikimedia Commons

public etiquette and wore a shirt, vest, and long trousers. They waded barefoot into the warm water, and Joseph showed Nelly "what fun it was to float on his back, and then use his arms to pull through the water while he kicked his feet to go faster." Nelly was sure she would sink if she lay on her back, so Joseph held her shirt tight in his hands, and she looked up at the sky hoping she would float like her father. But she was too stiff and fearful, so it didn't work.

Then she tried swimming forward on her stomach; this was much more fun, as long as Joseph held her. Her arms and legs churned in the water, creating lots of splashing. Her eyes burned from the salt, and she swallowed several mouthfuls. Swimming proved difficult, but in the days to come, Nelly practiced with her father and watched the American and Hawaiian children swim on their own. By the second week, Nelly, determined to succeed, swam without help for about fifteen feet and could float on her back. Joseph later mused to Malvina that sailing a ship around Cape Horn was a simple task compared to teaching Nelly to swim.[11]

The most meaningful activity during their stay in Lahaina was going to church each Sunday. They sang the familiar hymns, recited the prayers and responses, and listened to the gospel readings and sermons from the minister. Malvina also admitted in her letters to her sisters that she was excited to put on her best frocks with the latest starched sleeves, show off her new bonnets, and dress Nelly in the prettiest fashion when they all went to church. But the month was coming to a close, so their belongings had to be "gathered up from airing in the sweet-smelling breezes of the island" and repacked in the trunks, and the *Aurora* had to be loaded and inspected for the voyage south to the Marquesas Islands—a place on the globe that was the most remote from all continents.

One midday during the journey, while Malvina was calculating the ship's position after shooting the sun, she found, to her delight, that they were, at that moment, crossing the equator. She wrote down the degrees and proudly added her initials. But no whales had been sighted for several weeks, and the *Aurora* had only 980 barrels of oil stowed. Joseph planned to take on supplies at the island of Tuhuata in the Marquesas before sailing on to Tahiti. He had traded there before, when he was first mate to Captain Joy on the *President*, and he had learned how to safely conduct transactions among the quasi-civilized natives. Lotete, then the king, had been very eager to obtain knives, jackknives, strings of beads, and calico in trade for any amount of bananas, coconuts, oranges, limes, and sweet potatoes.

The island was now ruled by his nephew, King Maheono, who also was very interested in trade goods, including metal dinner plates that could be made into high status head ornaments. The new king was willing to send a hostage to the ship while trade was transacted, and he sent his son—the prince—whose rank was such that he had to be formally presented to Malvina, whose

Queen Pomare IV of Tahiti (1813–1877). *Queen Aimata Pomare became queen when she was only thirteen years old, but her 50-year reign was far from serene. Having had no children from her first marriage, she took a second husband with whom she had ten children. From 1843–1847 she fought a losing war with France for control of the Kingdom of Tahiti and was forced to accept annexation to France. However, she retained her title, royal properties, control over internal affairs, and legal access to the courts. Three of her children were accorded royal titles to Tahiti, Bora Bora, and Raiatea, while a son became President of the Tahitian High Court. Her palace was located outside Papeete.* – Pastel by Dawn Secord, in the style of Pierre Loti (Julien Viaud)

impression was that he "was a young man of handsome physical perfection, but hideously tattooed from head to foot on one side of his body only." Before he appeared on the quarterdeck for introductions, he dressed for the ceremony by putting on a short, calico shirt—and nothing else.

At the same time, Joseph took Nelly with him in the ship's boat to meet the king and present the trade goods. Two more boats were lowered and followed to provide transport for the food supplies.

Nelly later recalled the meeting:

> *My father said to me, 'Don't let go of my hand and don't say a word.' The natives surrounded us on the beach, they smoothed my hair, felt of my slender arms, wondered at my light skin but did not offer to snatch me away. As I was very little and not at all plump perhaps I was not very promising for a stew. Then the king called out to one of his men and a black chicken was brought forward and the king gave it to me as a gift for visiting his island. I held it tight as we returned to our ship. The king said good-bye to my father in French and my father replied. Back on board the* **Aurora** *the Prince bowed to my mother and left in his canoe for his home.*[12]

The next destination for the *Aurora* was Tahiti, in the Society Islands. This French port had skilled laborers who could tip the ship on its side to apply sheets of metal, so Joseph arranged for them to renew the copper sheathing on the hull of the *Aurora*, work that would take a week.

The most memorable part of this layover was the sheer paradise of Tahiti. The water in the harbor was so clear that "every detail of the coral was visible all over the bottom," and the Marshalls found that the smell of the tropical flowers, the woods, and fruits all blended into an intoxicating fragrance that hung heavy in the air. Joseph rented a French cottage and had meals sent in. The food was delicious; whereas the ship's cook's attempts to prepare breadfruit had never been appreciated, here on Tahiti it was baked on hot rocks for a very long time, served on banana leaves, and tasted like the

Tahiti Harbor, circa 1850. –Watercolor and ink from "Islands Seen by Ship Nautican," 1848–1853 by Susan Veeder, Nantucket Historical Association

finest soft, white potatoes. The young coconuts were tender, and the meat so soft it could be eaten with a spoon. Some meals were like those the natives ate and some were French. Pork was prepared in a variety of ways, and meals always included fresh fruit, like the ripe oranges hanging on the tree just outside the front door of the cottage.

One day, Malvina was invited to dine with the queen. While Joseph was ordering supplies and checking on the ship's overhaul, Malvina hired a small buggy and driver to take Nelly and her to the royal palace on the outskirts of Papeete.

As Nelly recorded the event later:

> *The queen received us in a garden where supper was being served for her, her ladies-in-waiting, and her guests. All sat on the ground and at each plate were baked bread-fruits wrapped in large leaves and various other kinds of fruits. The tall, stately queen wore a loose dress of some very thin material not unlike what used to be called a 'mother hubbard,' and she had flowers in her long, dark hair. She liked to see Americans and talked through an interpreter about mother's dress, particularly admiring a cameo breastpin in a very pretty setting and asked mother to give it to her. When mother replied that it was a present and so she could not give it away, the queen said, 'When you come back again bring [me] one just like it.'*[13]

During their visit with the queen, musicians entertained with instruments that resembled small banjos and were joined by others who added a distinctive beat with traditional drums. At the end of the day, whether because they thought Malvina and Nelly were British or simply to honor their own royalty, they played and sang *God Save the Queen*.

That night, Joseph and Malvina took chairs from the cottage and sat outside with Nelly between them. The night was warm, the sky was clear, and Malvina recited to Nelly a favorite quotation she had read long ago, when her own father had been at sea: "The majesty and power that holds every star in place—the glory of it all and how we are only a tiny speck—one of millions that go whirling about the heavens." Tahiti was a magical place of the greatest splendor, but tomorrow would be their last day before the long cruise across the Pacific to Peru. The pursuit of whales would continue.

House of Queen Pomare IV (right) alongside the Governor's House in Papeete, 1848.
– Illustration by Charles Giraud from *L'Illustration, Journal Universel*, No 278, June 24, 1848, Library of Congress

"A scene near Galapagos Islands, July 15, 1857." *Third mate on the* Aurora, *George E. Mills, illustrated the ship's journal with this patriotic whaling scene.* – New Bedford Whaling Museum

Chapter Nine

Mr. Hussey and Mr. Mills

George B. Hussey of Fall River, Massachusetts, and George Edgar Mills of Grafton, Vermont, were two young men working their way "up the ropes" in the demanding and dangerous vocation of whaling. While the labor of whaling was built on a fixed hierarchy, it also produced a sea-going brotherhood whose relationships were cemented by the rigors of facing common adversaries. The challenges of the working and living conditions on shipboard were exacerbated by long hours, unhealthy food, often-inadequate medical attention, isolation from the world at large, and monotony. And the imperative of hunting, killing, and processing a giant of all earth's creatures—the sperm whale—was ever present. Courage and strength were essential, for these were the men who drove the lances into their prey; these were the men who shouted with pride when spouting blood indicated a kill—"Chimney's on fire, chimney's on fire!"

It had been one hundred years since Yankee seamen first went forth in like pursuit, secure in the scriptural assurance that mankind had dominion over the beasts of the sea:

> *And God blessed Noah and his sons, and said unto them, "Be fruitful and multiply, and replenish the earth. And the fear of you, and the dread of you, shall be upon every beast of the earth, and upon every fowl of the air, upon all that moveth upon the earth, and upon all the fishes of the sea; into your hand are they delivered." Genesis 9:1–2 (KJV, 1810)*

Personal items used aboard the *Aurora*. *From left: Joseph Marshall's pocket copy of Emerson's essays, "Compensation and Experience"; his copy of Whittier's poems; his reading glasses; and the pin cushion that Malvina took on her two voyages.* – Photographs by Jann Gentry; objects from the Brown Family Collection

But Hussey and Mills, although devout Christians, appeared to approach their work with a more pragmatic view. Competitive and eager to succeed, they faced the hunt, however gruesome, with the same focus as those who hunted animals of the field and forest for food and skins. Yet, like men in other professions, they lived in two worlds—one of harsh inevitability and one of kindness and joy. They were mates to a captain they respected and whose family they treated with fondness. It is this latter world that stands in such contrast to the common perception of officers on a whaler, for both men not only enjoyed the company of Malvina and the working relationship they had with her, but they were also smitten with the precocious and ever curious Nelly.

The *Aurora* was now cruising in the open ocean, and the crew was optimistic about success in their new destination. No school lessons were scheduled for the day because Malvina was working with Joseph and Mr. Berry taking sightings, computing the ship's location, charting latitude and longitude, and recording the findings. Nelly sat on her usual perch atop a large coil of hawsers trying to push a piece of string through the holes in her colored beads. It just would not fit, and she was very disappointed because she wanted to make a necklace. There were tears welling up in her eyes when Mr. Hussey sat down beside her, took the string from her hands, and asked, "Do you know how to make a Cat's Cradle?" It took their four hands working together in steps to finally make the cradle. It was so much fun that they went on to make a Jacob's ladder with the same piece of string.

Mr. Hussey had been Nelly's chum since the first day on the *Aurora*, when he bowed from the waist and smiled at her as soon as they met. Although she did curtsy politely in response, she thought at the time that he might be teasing her—but in a friendly way. He was the youngest of the ship's mates, older than her brother, but much younger than her parents. Since he was often on the quarterdeck, Nelly was able to follow him around and often asked questions. He was surprised to find out how much she knew about the ship, the sails, and the lines. She often asked him to repeat the names of the sails and explain why they were raised and lowered at different times. Of particular interest to her was the huge compass housed just in front of the ship's wheel, and she had often asked her father to lift her up so she could see it, but was always denied and told it was an unsafe place for her to be. But Mr. Hussey devised a way that Nelly could look at the compass. He piled up three crates from the deck at the side of the housing and lifted Nelly up so she could stand on top of the crates to study the intricate face of one of the essential tools of the ship. It was exciting for Nelly to be up so high and look out at the deck from a different place. She imagined herself shouting out orders using the same words she heard every day—"lower away," "clear the decks," and "make all sail." But Mr. Hussey was very careful to explain that there were strict rules that must be obeyed and that she was permitted to climb on the crates only when he was with her. Nelly agreed.

In the weeks that followed, everyone was busy at ship's work. The mates planned and directed the painting of the bottom of the whale boats with "a mixture of yellow ocher and paint oil which dried like glue with a slick surface."[1] Joseph unrolled the canvas that he had purchased in Tahiti and assigned some of the sailors to repairing and patching sails, while others looked after the number of animals carried on board for provisions and cleaned up the refuse left on deck. There were "four goats three of which were lactating and one was a yearling, twenty chickens, eight pigs, and one bullock."

Malvina organized Nelly's lessons so that there were projects she could accomplish mostly by herself and thus allow her mother more time to care for James Boye, a Kanaka from Hawaii who was very sick. One of Nelly's major projects was to lay out and assemble a sample patchwork quilt for which Malvina had cut a variety of cloth pieces. It was painstaking and time-consuming work, but it was fun to wear a thimble on her finger as her mother did. It took her more than a month to stitch all the pieces together, so long that she had to be occasionally reminded of her commitment to a job that was meant to be an example of her ability. But no matter how difficult she found the task, she was proud of her accomplishment, even though the little quilt, still without finished edges, was not a complete success.

Personal items used aboard the *Aurora*. *From left: pastry cutters for hearts and rounds; whale oil lamp snuffer-trimmer with hearts—a wedding gift to Joseph and Malvina; and a handmade blanket pin for use on clothesline.* –Photographs by Jann Gentry; objects from the Brown Family Collection

The *Aurora's* cross-Pacific cruise from Tahiti to Callao, Peru, took it past Krusenstern Island in the Tuamotu Archipelago. At Chero, also known as Henderson Island, in the Pitcairn group, Mr. Berry and Mr. Mills went ashore to trade cloth, files, knives, tobacco, and an ax, for sweet potatoes, cabbages, taro, pumpkins, coconuts, wood, and a chance to refill the demi-johns with fresh water.[2] Soon afterwards, off the coast of Easter Island, a whale was sighted, taken, and processed for a total of 80 barrels of oil, but only a little later the captain encountered a difficult problem. When the *Aurora* got to Peru, Mr. Berry, the first mate, requested a discharge and asked to be relieved of duty. He had been a valuable first officer and both Joseph and Malvina tried to persuade him to stay on, but to no avail. Accordingly, after taking another whale off the northern coast of Chile and putting into Callao in October of 1858, Joseph signed the discharge for Mr. Berry, who left for Payta, took a steamer to Panama City, which was the newly opened Pacific terminus of the railroad across the Isthmus of Panama, and finally booked passage on a steamer to return home.

Joseph promptly announced the new officers: Mr. Hussey became first mate, Mr. Mills was advanced to second mate, Daniel Nevers, third mate, and Antoine Francis, fourth mate. Thereupon they sailed for

Patchwork quilt made by Helen Marshall as a school assignment aboard the Aurora. –Nantucket Historical Association

Tumbez to take on supplies and allow the crew to enjoy a brief liberty. Joseph purchased whale guns for the mates, so that they could replace the lances with this more modern equipment. Malvina deposited mail from the crew and collected the ship's mail, including letters, newspapers, a large tin container of bayberry leaves, and jars of cranberry preserves from Elizabeth and William. There was also a box for Nelly, but on this occasion, the "grand opening" would have to wait until they returned to the ship because the Marshalls had made plans to attend the open-air theater next to the marketplace, where a marionette show was scheduled. Judging from the size and makeup of the crowd, it became apparent that this performance was intended for adults as well as children. While Joseph and Malvina thoroughly enjoyed the performance, Nelly was dazzled and immediately began devising ways to put strings on her dolls so that she could make them dance and walk.

Back on the *Aurora*, after her father broke the soldered seal of the shipping crate, Nelly excitedly opened her box, in which she found a sun bonnet from her grandmother, a new slate to use for spelling and numbers from her brother, and a miniature rope bed and mattress from Auntie Elizabeth and Auntie Helen.[3] It made the whole day seem like Christmas, and her parents laughed quietly when they heard their seafaring daughter exclaim, "Clear the decks; I'm going to gather my things together and make them ship-shape." She put her newest doll, Nelly, on the mattress in the little bed and exulted, "Oh joy, the quilt just fits over her."

Meanwhile, the mates were engaged in familiarizing themselves with the operation of their new whale guns, and Joseph sent loaves of white yeast-bread to the fo'c's'le for the sailors.[4] Malvina offered a handful of bayberry leaves to anyone on board who wanted to sew a "remember pillow"—a fragrant reminder

"Shooting a whale with a shoulder gun," 1857. *The first successful shoulder guns were introduced in 1846, and by 1858, when Captain Marshall first used them on the* Aurora, *most whalers were using Robert Brown's lightweight gun patented in 1850. Brown streamlined his gun in 1855 in Honolulu, and they were readily available throughout Pacific ports. A bomb loaded into the gun was ignited by a flash of powder which also lit a fuse. The harpoon or lance was thrust into the whale and the fuse set off the explosion in about 10 seconds.* – Painting by whaler/artist Robert W. Weir Jr. aboard the Clara Bell, New Bedford Whaling Museum

of home—for their sea chests. But the next morning brought sad news and a solemn business to attend to: James Boye, the sailor Malvina had been caring for, had died, and the committal service was scheduled for sunrise. An anchor was dropped to keep the ship very still; and Nelly, who was not told of the loss but was kept occupied in the main cabin by Mr. Mills, never noticed a change in the ship's movement or regimen, nor did she later notice her mother's tears.

While Mr. Hussey had always been Nelly's favorite chum, Mr. Mills was also a special friend. He was fun to be with and had a big laugh, a wonderful voice for singing songs, and a set of colored pencils for drawing pictures. Sometimes he drew pictures of the sea and the whales being chased by the sailors in their boats, and sometimes he drew funny pictures of "monkeys and tortoises with smiles on their faces." When Mr. Mills was not on duty, he spent long hours writing in his journal and composing letters to loved ones at home. If his stateroom door was open, it meant that Nelly was welcome to visit. He allowed her to use discarded scraps of paper to draw pictures and occasionally lent her a colored pencil—sometimes two—to draw with. She stood right next to him and used a very small portion of the desk.

One day, when she asked about the little package on the desk's shelf, Mr. Mills was pleased to tell her about his friend, Frances, who had recently become his fiancé, and he unwrapped a framed daguerreotype to show Nelly the pretty face of a lady with a bonnet. She did not know what a fiancé was but sensed that she had asked enough questions for the moment.

So Mr. Mills returned to his journal:

> *There are five ships in sight and spoke to Capt. Starbuck of the Bark* Islander. *Got a whale 35 barrels, Capt.*

The whaling ship *Islander*, circa 1860. *The 347-ton ship, built in 1856 in Fairhaven for Nantucket merchant Matthew Crosby, was always known to be rigged as a bark. However, in this painting, it is rigged as a ship, trying out oil in the South Pacific.* – Unknown artist, Nantucket Historical Association

Steward and his Lady came aboard from the ship Mary and Susan, *took whales today—three and one was 100 barrels. Our boats have been stove three times this month no one is hurt and we have 950 barrels total. Good fortune, I got letters from the box on Charles Island. I killed a blackfish with my gun. Gammed again with Capt. Swain from Fairhaven. Six little pigs were born this morning.*[5]

Later that week, Nelly developed a sore throat and was put to bed with a cold and fever. Malvina sat with her and read stories, but Nelly was fretful and wanted to play on deck instead—that is, until Mr. Mills arrived with a little bag of horehound drops for her throat and a secret bundle under his arm: one of the new piglets for Nelly to see. It was wrapped in a piece of sailcloth and was very small, and although Nelly offered to care for the piglet, Malvina explained that the newborn had to remain with its mother.

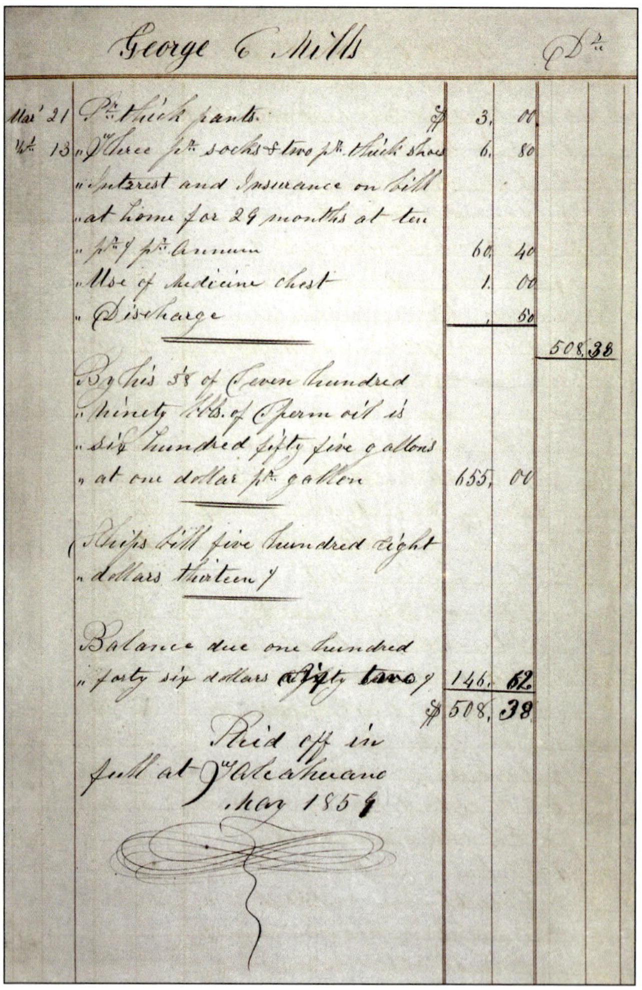
George C. Mills Dr.

Mar 21	Pr thick pants	3.00	
" 13	Three pr socks & two pr thick shoes	6.80	
	Interest and Insurance on bill at home for 29 months at ten pr % pr annum	60.40	
	Use of medicine chest	1.00	
	Discharge	.50	
			508.38
	By his 58 of Seven hundred ninety bbls of Sperm oil is six hundred fifty five gallons at one dollar pr gallon	655.00	
	Ship's bill five hundred eight dollars thirteen		
	Balance due one hundred forty six dollars sixty two	146.62	
		$508.38	

Paid off in full at Talcahuano May 1859

Account journal of bark *Aurora*, 1856–1861. *Captain Marshall's journal of disbursements and accounts shows that thirty-three crew members were engaged at some point on the five-year voyage. Of the original crew that left New Bedford in 1856, only nine returned. Among those who did not were ten deserters, twelve discharges, and at least two deaths. George Mills was among the discharged, leaving the ship in Talcahuano, Chile in March 1859, and earning himself $508.38, after expenses, for two-and-a-half years' work as second mate.* – Brown University Library

Nelly recovered quickly and began planning what she wanted to do next. Having watched Mr. Hussey begin the dog watch after supper, she reasoned that she would be capable of standing the shorter watch with him. Her parents said that she was too young, but Nelly persisted; and when Mr. Hussey agreed to have her join him—as Nelly knew he would—Joseph and Malvina relented. Soon she was clad in her pea coat, wool hat, and leather boots, and except for the petticoat and ruffled skirt that hung beneath her jacket, she looked like a miniature sailor.

Up and down she paced the deck trying to keep step with Mr. Hussey. She took long steps and followed right behind until the air grew chilly and she got drowsy. In fact, she became too sleepy to keep her eyes open and had to be taken to the stateroom and put to bed. But the next day, Mr. Hussey complimented her for standing watch with him, so she felt that she had been a success.

Nelly was still curious about another matter, though. She kept thinking about Mr. Mills and the picture of his friend, Frances. So she asked Mr. Hussey about his desk, but he was very abrupt and stern, telling her that his desk was private and not for others to see. He added that friends were obliged to respect each other's personal things. She said she understood, but she really did not, and rather than going away, her curiosity increased to the point she couldn't ignore it any longer:

The next time the boats were all off after whales, she watched [for] her chance and opened the door and climbed up to the desk to explore what mystery it contained. All she found were some papers carefully tied up, and the big log book. But in the left hand corner was a small, pasteboard box. Inside was a daguerreotype of a young lady with a sweet face wearing a white bonnet with rosebuds on the side.[6]

In the last month of 1858, Joseph planned an aggressive push for whales. More than half of the oil taken by the *Aurora* had been brokered in Talcahuano, and the Captain's goal was to refill the ship's hold with additional barrels of oil to insure a full cargo on its return to New Bedford in order to maximize the voyage's financial success. He had three good and very specific reasons. First, this would be his last voyage before retiring; second, it was obvious to experienced whalers that the population of whales was diminishing and the worldwide market for whale oil was waning due to the recent discovery of petroleum; and third, he knew that the attitude and morale of the crew, most of whom were in their third year of service, would not improve with time, and good replacements were becoming very hard to find.

Joseph had already put parts of his plan in motion when he bought the whale guns, which made it easier to dispatch a harpooned whale and doubled up the whale watch high in the crow's nests. Even more important, perhaps, he carefully studied the log to find where the most whales had been taken and how long it took to get from one area to another. The result was that he decided to cruise off the coasts of Chile and Peru, ranging only as far west as the Galapagos and Juan Fernandez Islands. He noticed by his whale count that the greatest number of whales had been taken in the vicinity of Masafuero in the Juan Fernandez group and the greatest number of sperms was taken between the Galapagos and Callao. The information thus gathered was not particularly new, but the specific data were considerably more convincing than inexact recollection. Sailing south toward the indicated whale grounds, the *Aurora* took two whales on the way and a third just

Whale chart by Matthew F. Maury, U.S. Navy cartographer, 1851. *The map identified water temperatures, weather, whale sightings, and whale killings. The map is divided into sectors that use color and pictorial symbols to show seasonal variations and to identify whales by species.*[7] – Library of Congress

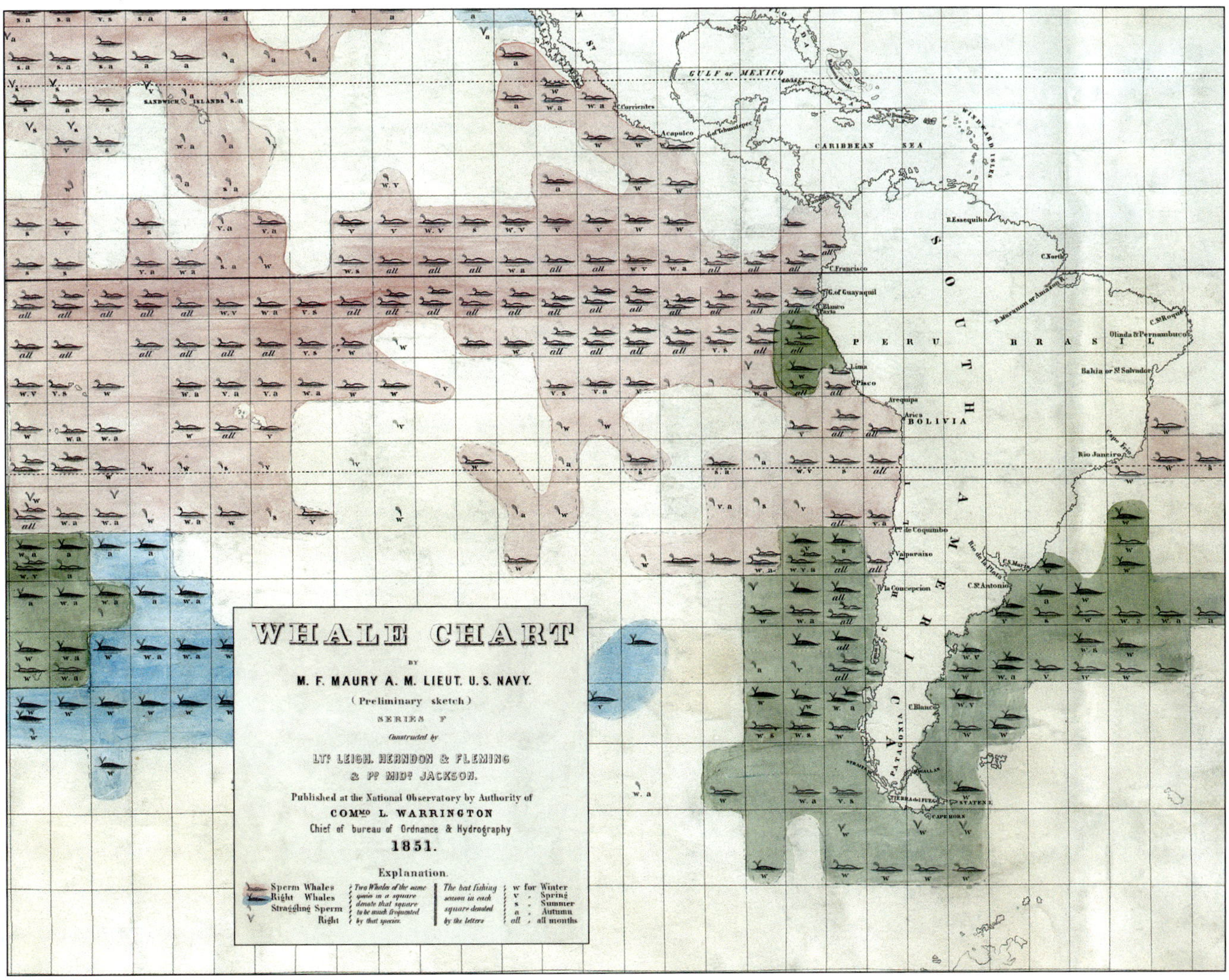

View of Callao, Peru, 1836.[8] *By 1845 the port had grown to become a major whaling supply destination, and included a ship chandlery and counting house owned by William Crosby's brothers, Matthew and Sylvanus.* – Etching by Barthelemy Lauvergne, National Library of Australia

west of Callao. The plan appeared to be working. The sailors chalked it up to seaman's "greasy luck."

Nelly celebrated her birthday at sea, for there was neither time nor need to go into port. The supplies on board were ample, white flour had been set aside to make a cake, and the cook had enough tiny peppermints to spell her name on top. Joseph and Malvina gave Nelly a mahogany-framed primping mirror—for she was growing up to be a pretty girl. And one of the sailors presented her with a very special gift that he had made just for her—a blue painted jump rope with carved whale bone handles ending in large, elaborately fashioned Turk's heads. It was beautiful, and it was the perfect gift for the girl everyone knew had the energy to use it.

But Nelly was still intent on being a sailor. She studied the names of the sails, learned how to read the compass, and with her mother's help, studied the big maps in the main cabin. She repeatedly asked to climb up into the boat where Mr. Hussey and Mr. Mills "took the sun" with the quadrant but was each time firmly refused. One day, however, with no one looking, she took her chance, climbing and then pulling herself up until she reached the edge of the whale boat. But she slipped and fell and would have fallen overboard between the boat and the ship had Mr. Hussey not seen her in time to catch her by the dress. They were both thoroughly frightened, and Nelly begged the mate not to tell her father and mother. He promised—and always kept that promise.[9]

What they did not know, and what Nelly was soon to find out, was that Joseph had seen the event from a distance. He walked her to the quarterdeck, dismissed all who were there, and in private, severely reprimanded his daughter. He also instructed her to say no more of this to anyone—that her mother must not hear of it—and to never disobey the shipboard rules again. Nelly was so scared by what had happened and alarmed by her father's stern words that she dissolved into tears but could not tell anyone why. It so happened, however, that Mr. Mills was a third witness to Nelly's accident, and he recorded it thus in his journal:

Nelly, Dearest Nelly
Why shed those pearly tears
Why pout thy red and cherry lips
And swell those pretty eyes
Thy Pappa loves thee dearly
Yet fore thy good gives pain
Now go and tell him kindly
You'll never do so again

Now hear as you cling to him
T' was sad for him to hear
He loves thee Nelly dearly
Then pray those tear drops hide
We all feel sad and lonely
When your cheery voice is gone
So cheer up pretty Nelly
And have with your sorrow alone

Nelly jumping rope on deck of the *Aurora*. *Joseph and Malvina Marshall were concerned about the limited exercise available to their daughter aboard ship so they encouraged her to jump rope. Made by a sailor for Nelly, the rope was painted blue and the handles were hand-carved.* – Pastel by Dawn Secord

The Tumbez River in Peru, 1859. *Tumbez was a favorite port of call for American whalers. The river water was sweet and provisions such as potatoes, plantains, oranges, fresh beef, and wood were ample and cheap. The river also provided a passageway to the Somontez plantation.* – *Frank Leslie's New Family Magazine*, Vol. 5, No. 4, October 1859, Library of Congress

Chapter Ten

Homeward Bound: Seaquake, Goodbye to Somontez, and Civil War

Joseph sat alone in his cabin among the host of ledgers, logs, and maps, engaged in serious introspection of his personal life. He regretted having spent so much of the time at sea when his son, Joseph was growing up at home; now he began to question the wisdom of taking Nelly to sea while she was just a baby and having her on board when she was so active and full of risk-taking adventure. Her near-accident led him to profoundly question his ability to make wise family decisions, and he was keenly aware that Malvina's quite-understandable fear of her child's falling overboard had been reinforced by a disturbing nightmare with the same scenario. Surely, she did not need to be further burdened by what had occurred. But his responsibility for the well-being of his family was not his only concern, as he was also responsible for the welfare of the ship, its valuable cargo, and all on board. He must, he decided, be resolute in all matters and chart a final course for home. While cruising between Payta and Valparaiso early in 1859, the *Aurora* added enough to her cargo that by spring she could boast 400 barrels of sperm oil as well as a goodly supply of whale bone and teeth. As a result, the ship, the crew, the officers, and the Marshall family had taken on a persistent environment of grime, grease, and smoke. To those who would profit from the voyage, the horrendous odor emanating from the try works was the smell of success. To Malvina and Nelly, however, it became a putrid and ever-present reality.

Finally, the jagged volcanic peaks, conical hills, and barren black rocks of the Galapagos hove into view, and when, at last, the *Aurora* dropped anchor in a sheltered bay, a thorough process of cleaning and painting began. Soon only the gray sails were reminders of the weeks of fire and smoke, and two boatloads of sailors were sent ashore to capture and return with a "goodly number of turpin [tortoises] for the next months' rations" while the ship was en route to Payta and beyond. Many of these giant land tortoises were so large that it took two men to lift one into the boat, but the meat was so delicious and could be prepared in so many different ways that it became a popular main course. An additional and more practical value was that tortoises could live for several months without food or water. Great exaggerations were often recorded about the viability of the creatures for food use, but in truth, the quality of the meat was not very palatable after several months, and sometimes the remnants were tossed into the try pots.

Two whales were taken without incident, and when all the work was done, small cash rewards and squares of tobacco were doled out to the sailors—this time before they went ashore in Payta for liberty. In addition, Joseph and Malvina presented the mates with practical gifts of clothing from the ship stores. Mr. Mills was especially pleased with his new hat and recorded in his journal that he would "wear it in the captain's honor." After a brief two days in port, the *Aurora* sailed for Tumbez, where Joseph arranged for the loading of provisions, paint, and wood, while Malvina and Nelly walked through the marketplace buying exotic fruits for the officers' table, oranges for the crew, and unique trade goods for gifts to take home. But before the family began the trip upriver to their planned farewell visit to the Somontez family, Nelly became ill from eating too much fruit, and Mr. Hussey was enlisted to escort Malvina and a very disappointed Nelly back to the ship.

Joseph hired a horse and left on the river road toward the plantation. The Somontez family was pleased to receive him but saddened to hear it was to be his last visit. They enjoyed a special meal of roasted goat and reminisced about their years of friendship. Then, as Joseph rose to retire for the night, he abruptly fell to the floor, first on his knees and then on his side. When Señor Somontez and his servant tried to help him up, they found that Joseph was not fully conscious. They

View of the marketplace in Payta, Peru. – Painting and lithograph by Barthelemy Lauvergne, artist on the voyage of *La Bonite* 1836–1837, Wikimedia Commons

carried him to the bedchamber that had been prepared for the visit, and determined that he was suffering from a very high fever and was unaware of his surroundings. The Señora herself directed the servants in caring for him until help arrived:

His outer clothing was removed, cold cloths were applied to his face, arms, and legs, and cold tea was held to his lips to drink. When he did not improve in an hour, Somontez sent for the doctor and dispatched his private boat to [send word to] Malvina and request her immediate presence.

Since the captain was not expected back on board until the next day, the watch on the *Aurora* was somewhat surprised to see a harbor boat approaching the ship so late at night. A messenger came aboard requesting to talk with Mrs. Marshall and "to deliver a note into her hands." Somontez had been sensitive to the fact that this news might be alarming for the captain's wife, and therefore inappropriate for the officers to hear about before the facts of the matter were known. He also recognized the likelihood of misunderstandings when people depended on translations. Therefore, the message simply explained that the captain had taken sick at the plantation and requested his wife's presence while he recovered. It also said that Señor Somontez's boat was waiting on the shore. No further instructions were necessary, since Mr. Hussey had already been placed in command of the *Aurora* during the captain's absence. Malvina calmly yet quickly packed necessary personal items for herself and Joseph and accompanied the messenger to the dock, where the river boat was waiting. She later recalled:

> *Somontez sent a most sumptuous boat with ten rowers [to provide for me]. Never will I forget the trip up the river in silence, the warm night, the brilliant stars, the rowers in perfect unison striving their best to get there as fast as possible. It was near midnight when we arrived. Was I frightened? I knew I would have been no safer if the whole United States Navy had escorted me.*[1]

Yet Malvina was not prepared for Joseph's appearance or the serious condition that it indicated. When she took his hand, it was hot and sweaty, and he intermittently shook as if suffering from chills and pain. The doctor had left instructions for offering liquids and treating the chills that would arise if the fever broke. He also reported that Joseph's pulse was weak and slow—a detail that concerned Malvina more than any other, for her own father had died from "complications of the heart" in Pernambuco on his way home from a four-year voyage to the Pacific.

But, for now, all she could do was to give her husband the care she thought was best. She propped up his head and talked to him about the familiar business of the *Aurora* while she bathed him and covered him with clean linens. Before morning, Joseph's fever broke, he regained consciousness, and he spoke briefly to Malvina before beginning to vomit. When the doctor returned, he advised bed rest, tea, and whatever food was possible to keep down. After several days of very slow improvement, Somontez and Malvina decided that recovery would be easier with the Marshalls accommodated in a private cottage, and that one of the servants would continue to attend to their needs. Malvina sent a letter to Mr. Mills, requesting him to deliver Nelly to her parents at the plantation and to return to the ship with the captain's written orders for Mr. Hussey.

Nelly would later describe the several days before they arrived at the plantation as a time of "greater freedom on the ship, interesting travel on a river boat, and learning to fish with a pole." Among other things, she was allowed to enter the ship's galley and talk to the cook as long as there was no fire or cooking in process. She particularly liked watching him clean out the turpin shells, and although these were usually given to the sailors to turn into combs, hair pins, and enhancements for lady's bureau sets, five of the largest ones were set aside this time and stored carefully between decks for the cook's future use.

As she got ready to join her parents at the plantation, the steward helped Nelly pack a bag for her trip, and Mr. Hussey surprised her by preparing the gamming chair for her descent down to the boat. Mr. Mills and Nelly traveled up river reciting poems and singing songs in the heat of the January day. But, when they arrived at the plantation, the sight of Joseph was a puzzle to Nelly and a concern for Mr. Mills, for he was still confined to bed and his usually tanned face was pale. The captain met with his mate and handed him a note with instructions for a month's cruise off the Peruvian coast without liberty at any port, but these were written and signed by Malvina. On his return trip to the ship, Mr. Mills gave considerable thought as to how he would describe the captain's illness to the crew, lest the report cause undue anxiety among them.

Nelly had been only somewhat curious about her father being in bed in the middle of the afternoon, but she genuinely hoped his headache would go away soon so they could take a walk to visit the sheep and goats on the plantation. As it turned out, she would have to wait almost a month for that walk.

The captain's recovery was slow, but as the weeks passed, Joseph regained strength, adding short walks to his daily regimen, playing checkers with Nelly, and enjoying chess with Somontez. Finally, Malvina was confident that a return to the ship would enhance the recuperative process. The Marshalls had been the grateful recipients of Señor and Señora Somontez's "untold generosity and thoughtfulness," and they found it hard to leave, knowing that they could never adequately repay such kindness.

The return to the *Aurora* in the third week of February brought joy, but also misgivings. The Marshalls were once again surrounded by the familiar elements of their home at sea, yet the crew was apprehensive about the captain's fragile appearance since his illness and began to question his ability to get the ship and cargo ready for the long trip home.

Then, on April 3, Mr. Mills was at the helm on a course south to the Juan Fernandez Islands when he saw a wide swath of ocean water move at high speed directly away from the bow of the ship; seconds later, the ship dropped into the resulting void and was severely shaken by the impact. Sailors appeared on deck from every part of the ship, staring in disbelief and fear. Captain Marshall arrived on deck as the third wave of the seaquake hit,[2] only to witness fourth and fifth plunges and violent shakes that cracked the brick and masonry try works and splashed the water out of the protecting moat. After what seemed like long moments of silence, the captain explained what had occurred and ordered a complete inspection of the ship. Fortunately, nothing else was damaged, all cargo was secure, and soon five skilled craftsmen were hard at work rebuilding the try works.

Still, some members of the crew began talking about omens and strange coincidences, especially since the only damage to the ship was confined to where dead whales were processed. And their lingering concern about the captain's health only fueled their superstition. But this sort of gossip ended abruptly when another call was heard from the crow's nest and the boats were lowered away.

Soon after the unusual occurrence of the sea quake, Joseph had become aware that George Mills was not able to work to full capacity. He had previously been ill with various ailments, and now it was a severe case of skin "boils."[3] Malvina also noticed that he demonstrated symptoms of "ship fever"—that malady of depression common to men at sea. So it came as no surprise when Mills requested a leave at the next port to seek medical care, but after discussing the situation, the captain and Mills agreed that he would be paid in full for time served and discharged. With some reluctance over losing a once capable mate, while also facing the reality of a crew three men short, Joseph put into Talcahuano to bid Mills good-bye and arrange for supplies and new recruits. The *Aurora* then made sail immediately for the whaling grounds west of Valparaiso.

By October there had been no "greasy luck," the morale of the officers and crew was at its lowest ebb, and Joseph and Mr. Hussey began to notice a sluggish reaction to the ship's helm when coming about. Upon careful inspection, second mate Daniel Nevers,[4] the most experienced man on board except for the captain, found a leak in the hull.[5] But Joseph and his officers decided that, with bilge-pumps at the ready and sufficient ballast provided by the barrels of oil in the hold, it would be safe to proceed directly to Callao, Peru and the ship chandlery owned by William Crosby's brothers, Matthew Jr. and Sylvanus. Though not the closest docking facility, it was American, it was family, it was completely trustworthy, and the company's agents could handle the required bribes to port officials so that the *Aurora* could be taken in for repairs in a timely manner.

The damage to the ship was a significant setback, especially since almost a full cargo of oil was on board and the officers and crew were hoping to be homeward bound. In addition, the superstitious inclination that haunts all sailors raised a nagging fear: what might be next after the captain's temporary but debilitating illness, the quake at sea, and now a damaged ship? Joseph, meanwhile, faced the complex challenges of keeping the crew together while the *Aurora* was in dry dock with its cargo removed and making up for the cost of repairs and months lost from hunting whales. This was his last voyage and it had to be profitable.

His plan was a good one, and he knew he must make it work. The crew would be rotated as "guards" on the *Aurora*, under the command of a mate for each watch. This would insure the safety of the ship during repairs and keep the men employed. The Crosby Company[6] would become agents for the sale of the oil, and when the ship was ready to sail again, Joseph would extend the voyage until a new cargo of oil was safely stowed in the hold. Finally, he would offer cash rewards to be added to each man's lay at the end of the voyage.

So, it was with some clarity of solutions and purpose that Joseph sailed into the port city of Callao. Malvina and Nelly were on deck to watch all the arrival activities and were already packed to move to a hotel, until a house was found for their accommodations. The Marshalls were also looking forward to a reunion with family members and catching up on all the recent Nantucket news.

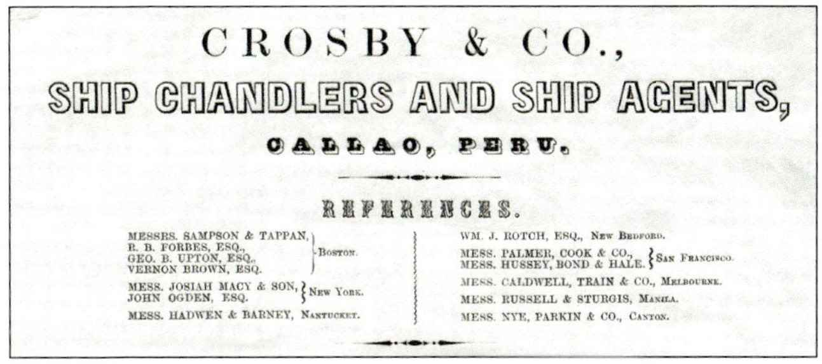

CROSBY & CO.,

SHIP CHANDLERS AND SHIP AGENTS,

CALLAO, PERU.

REFERENCES.

MESSRS. SAMPSON & TAPPAN, R. B. FORBES, ESQ., GEO. B. UPTON, ESQ., VERNON BROWN, ESQ. — BOSTON.
MESS. JOSIAH MACY & SON, JOHN OGDEN, ESQ. — NEW YORK.
MESS. HADWEN & BARNEY, NANTUCKET.
WM. J. ROTCH, ESQ., NEW BEDFORD.
MESS. PALMER, COOK & CO., MESS. HUSSEY, BOND & HALE. — SAN FRANCISCO.
MESS. CALDWELL, TRAIN & CO., MELBOURNE.
MESS. RUSSELL & STURGIS, MANILA.
MESS. NYE, PARKIN & CO., CANTON.

Masthead of the advertisement letter *soliciting new business for the firm of Crosby & Company in Callao, Peru, 1853.* –Nantucket Historical Association

By Christmas, they had moved into a house closer to the neighboring city of Lima. It was a comfortable place with a fenced yard in the back and plenty of room for Nelly to play. Joseph bought her a metal hoop and showed her how to use a stick to keep it rolling, not an easy skill, but one Nelly practiced as best she could. Then one day, Malvina brought home some flowers from the city marketplace and showed Nelly how to fill clay pots with soil from the yard and plant a flower in each one. Nelly enjoyed digging in the yard and getting her hands in the soil—and she really liked finding worms that wiggled when she touched them. With so many interesting things to do in the yard, she was not happy about having to do her schoolwork in the house. But Malvina kept regular hours each morning for

"Callao: The Harbour of Lima," mid 19th century. *The primitive watercolor painting in the collection of the New Bedford Whaling Museum depicts ships at anchor in Callao Bay. At center, the twin spires of the Cathedral Basilica of Lima can be seen, and at far right, the fortress Real Felipe is visible.*

lessons in arithmetic, French, history, and geography. She also bought white linen tea towels and bright red embroidery thread,[7] and taught Nelly how to cross-stitch her initials above the hems.

Malvina took Nelly to the magnificent white cathedral in Lima, to the museums, and to a park with statues of men on horses. They both enjoyed shopping in bookstores. Malvina bought a Spanish language book, and Nelly selected several story books written in English. On some days, they wrote letters to family at home—Nelly to Joseph W. and her grandmothers, Malvina to her sisters. And it was particularly pleasing to be close to the American consulate so they could receive mail without waiting for months. The maid in their house was surprised that Malvina liked to cook for the family. When Joseph was at home for the midday meal, it was often a repast that Malvina prepared "American style," and if there were guests, Malvina planned the menu and did most of the cooking.

However, most of Joseph's time was spent observing the work being done on the *Aurora* and overseeing the work of his officers and crew. With the help of the Crosbys, he also began looking for an additional mate and several experienced seamen. He had to use careful judgement in his selection of men, and he needed the best in order to capture as many whales as possible in the least amount of time. The problem was—qualified seamen were scarce at that time.

Yet on January 10, 1860, the fully repaired and newly painted *Aurora* raised its new white sails and headed out to sea with a full crew. The course was set for the Galapagos Islands, and despite two consecutive squalls that pelted the ship with fierce rain, the outlying islands hove into view a month later. Whales were seen but could not be taken. And after Joseph remained in the area hoping for an annual migration that never materialized, they put into Charles Island for a mail drop and boarded enough tortoises for the spring menus. Finally, their luck turned, and two large sperm whales were taken, followed by another one a week later. The crew was so busy that the cooper had to enlist the help of the steward to keep up with the need for more barrels, and the cook was working around the clock to keep plenty of food ready for the tired and hungry seamen. The real test of a skilled crew had just been passed: they knew how to work together. Now, Joseph needed a plan to keep his crew together.

All the men presently on the *Aurora* had been at sea for at least three years, some for more than four, and it was all too easy to find temptation while on liberty in the major ports of Peru and Chile. Joseph also knew how boredom and lack of work created problems. He poured over all his charts on whale kills, location, and time of year, charting his courses accordingly. He then outlined more obscure settlements—preferably islands, where he could purchase or trade for supplies, and at the same time, give his men time on shore. Then he gave away any pieces of nonessential wood, metal, turpin shells, and whale teeth and encouraged the crew to make gifts to take home. Finally, he took advantage of every possible gam, for visiting other ships provided a welcome source of sociability, news, and exchanges of mail.

More whales were taken off the coast of Valparaiso, and when the ship was clean and free from activity, several sailors began a project using pieces of wood discarded by the carpenter. They kept it covered with a tarpaulin when they were not working on it, though apparently the captain knew what it was. By the third week of November—the very day of Nelly's birthday—the tarp was removed to display a miniature dory with two oars and one tiny sail. And it had one thwart amidship just the right size for the smallest sailor on board.[8]

A month later, the *Aurora* gammed with a whaler from Fairhaven, and it was learned that Abraham Lincoln had been elected president of the United States. The Marshalls also received mail from home, and in the months-old newspaper accounts, Malvina read about the presidential campaign that had pitted Lincoln against Stephen A. Douglas, John Breckinridge, and John Bell. She called to Joseph's attention the editorials and political critiques by journalists and clergy, and read with great interest her sisters' comments on the unique theological positions of their local ministers. Rebecca had sent all her *Springfield* [Massachusetts] *Republican* issues with the Timothy Titcomb letters; Harriet had forwarded Theodore Parker's editorials in the new, erudite *Boston Herald*; and Helen had sent a *Banner of Light* with collections of Henry Ward Beecher's sermons. The burning topics of the day, of course, were Lincoln and antislavery. But it was not until a gam in the new year with several ships from New York that Joseph and Malvina learned that South Carolina had seceded from the Union and six more states were due to follow. The consequences of these developments for those hunting whales off the coast of South America would, before long, be both clear and worrisome, for surely war would soon follow.

Joseph revised his itinerary and set a course for Talcahuano. Good fortune smiled, for on the way, the *Aurora* took on 50 barrels of oil to meet the captain's minimum goal. He then prepared a detailed letter for Elizabeth and William, informing them of the timeline for the homeward voyage. They would leave Talcahuano in February, stopping only for supplies and mail. He was fully aware of the potential risk of giving the crew liberty in Chile, for he could not afford to lose even one man before heading for the always fearsome transit of Cape Horn.

But by the time they arrived at the port it was late March, and the United States Consul and the incoming mail brought warnings. The most serious concern was the fact that Lincoln's inauguration had created a final split between the Union and the Confederacy, and while the new president immediately ordered a blockade to prevent the South from carrying on vital trade, the Confederacy retaliated with privateers operating under Letters of Marque authorizing them to capture and burn Union shipping vessels of any kind—and unarmed whalers were a primary target. In addition, the north-bound route would lead them along the 3,400-mile coast of Brazil—whose government was sympathetic to the Confederacy—and then through the gauntlet of Caribbean islands and past the now dangerous southern ports of Savannah, Charleston, and Norfolk.[9]

Although Joseph's plan to keep his crew together had merit, it was clear that the men needed and deserved rest and liberty before the long and arduous trip to New Bedford. So he delivered to all hands a lecture bordering on a sermon about duty and responsibility shortly before the *Aurora* strategically docked at Talcahuano only after all other ships had departed. The first order of business for Malvina was to mail an urgent letter to Elizabeth with instructions to purchase a dress and bonnet in the latest fashion and have it delivered to the *Aurora* as soon as it docked in New Bedford. She made it very clear that she "would not leave the ship till she could put on something more suitable than what was in her wardrobe after nearly five years."

The *Aurora* was well over a month late on its previously planned schedule when it cautiously sailed well

Destroying whaleships, 1865. *The Confederate States warship* Shenandoah *(far right) was responsible for the destruction of the greatest number of whaling ships during the American Civil War.* – Lithograph from the U. S. Naval Historical Center, from a painting titled "Destruction of Whale Ships off Cape Thaddeus Arctic Ocean," by Benjamin Russell

clear of the rock-strewn coast of Chiloé Island on the way to Cape Horn. The weather turned cold and huge flocks of birds filled the gray sky below the threatening clouds. The cook, often "with his nose attuned to the mood of the crew as well as to their appetites," planned a "good-bye to the Pacific" celebration. With the captain's permission, on the first afternoon of clear skies, he called all hands to the deck for a dinner of tortoise stew with carrots and onions, corn meal biscuits, and plum duff. Then he enlisted several sailors to collect wood scraps from the carpenter's stall, soak them in whale oil, and distribute them equally in the bottom of the five inverted turpin shells—one for each year of the voyage—which he had prepared two months before. The sailors loaded the shells in a whale boat, lowered the boat into the ocean, and lit the fuel before setting the "turpin boats" adrift on the sea as night began to fall. Nelly stood next to her friend Mr. Hussey at the starboard rail and watched with fascination until the last of the five twinkling lights disappeared into the darkening waves of the Pacific.[10]

Although Malvina and Joseph stood together on the quarterdeck, Joseph was alone with his thoughts. He wondered whether the friends they'd shared ten years ago would remember them and still care for them. He wondered whether his letters to family members had been sufficient to keep bonds of love alive over the course of the two voyages. He wished that he "could have taken giant wings to carry Malvina and Helen with him and [also] be with friends and family before the sun sinks in the far west at night." Had he denied them a real home? As Joseph wrestled with these melancholy questions, he also addressed his very real concerns for the future. It had been thirty-two years since he first shipped out at the age of seventeen; he was weary of the sea. The economy of Nantucket was now in serious straits, and he was returning without a plan for employment. Would his savings and investments be enough to provide for his family? And he was returning to a country at war. But one thing he did know: his complete devotion to Malvina, "his best half," was stronger than any other on earth.[11]

Captain Marshall's Victorian black horsehair stateroom chair.
– Drawing by Dawn Secord, Thomas W. Hall III and Katherine Moore Hall collection

Noticing what looked like sadness in Joseph's face, Malvina began to remind him of the recent letters from Elizabeth and William. They not only reflected the reality of the Crosbys' new life at Bloomingdale Farm—far from the halcyon days of One Pleasant Street—but also their determined efforts to be farmers. And the stories of "William learning how to raise potatoes and Elizabeth driving her own poultry, eggs, and butter down the turnpike to market in town" were written to encourage laughter.

Malvina had her own thoughts about going home. She would return to the circle of her family, albeit a smaller circle than it once had been. Rebecca and Bobby Mitchell had long ago moved to Holyoke with their children; Helen and Joseph Belcher lived in Providence; and Harriet, the last to leave the nest, had recently married Boston barrister, John Locke, and lived in that city on the Commons. Seth Jr. and Linda had left Boston and were relocating to California. But her mother was still at 46 Fair Street, and in the winter months, when Elizabeth and William returned from 'Sconset, the house was filled with activities. Mary and her husband Henry Plaskett, who recently retired from the China Trade, were on island with their young daughter, Elizabeth, and other cousins were everywhere. And that was just the Pinkham side of the family. Although Nelly was happy with life the way it was, she would have plenty of young Marshall cousins to play with at home. And in addition to both grandmothers waiting with hugs and kisses and a cherished brother, Joseph, who would be home between his voyages to India and Asia, Nelly had her closest friend, Mr. Hussey who was always ready to play a game or tell her an adventure story. She knew she would be going to school, but that would be months away. And after almost five years at sea, her memories of Nantucket had faded. She wondered "if the children wore pea coats with gingham patches and whether she was equal to them in her studies of French and arithmetic."

As the *Aurora* entered the cold and stormy climate of Cape Horn, Joseph was aware as never before that the seamanship of his officers and crew was of the highest quality. He had written to William, "Just one year ago I kept changing in the hope of getting good men and have at last succeeded. The difficult course ahead will be handled well and I am fortunate for that." By April 15, they had cleared the Horn; by May 1, the Falklands were in sight, and they were only a month away from the tropical climate of the West Indies. The *Aurora* made one very brief and cautious stop for water and supplies on the Island of Ferdinand de Orsa on June 5, and at three o'clock in the afternoon, left for home.

From that point on, double lookouts were posted for any sign of privateers, and soon the ship entered the sixty-mile-wide Gulf Stream, staying on the easternmost edge, as far away from land as possible. Then the captain ordered the try works dismantled and thrown overboard to lighten the ship and speed it homeward. The third mate recorded "No observations for two days—sailed by dead reckoning, and last night we took the NE trades with hopes they will last and give us a good push well north." Ship debris was sighted floating on the waves in several places off the coast of Florida, but no craft of any kind appeared until a large steamer crossed the *Aurora's* path off the coast of Georgia. It turned out to be part of the Union blockade, and the crew of the whaler was much relieved to see the stars and stripes flying from the ship's jackstaff.

"Hove to for a Pilot," 1860. *Marine artist William Bradford rendered this scene of a homeward bound whaleship giving way to a pilot schooner that will guide her into New Bedford. Here, she passes through Vineyard Sound, round the Gay Head cliffs, before crossing Buzzards Bay.* –New Bedford Whaling Museum

But every hope for a quick passage home was soon dashed, for when they sighted Cape Hatteras and headed further out to sea for safety, the winds tapered off and the sails fell slack. To keep the men busy, Joseph ordered a complete cleaning of the ship in preparation for new paint. He wished to have everything in the best possible order before docking in New Bedford. And the men, although discouraged by the lack of wind and concerned about armed blockade runners operating out of Norfolk and the Chesapeake, began to lay out their best clothes, cut their hair, and shave their beards. Malvina set to work tailoring Joseph's finest suit so

Whaleships arrive at New Bedford, 1870. *The bark* Massachusetts, *her sheets drying in the wind and her bounty of oil spread along Rotch's North Wharf, had just returned from a successful Pacific voyage. At center (wearing derby) is local merchant and ship owner W. R. Wing accompanied by his daughters, Helen and Clara. Being seen around town in the 19th century—even on the working waterfront—was always a dress-up affair.* – Photograph by Stephen Adams, Spinner Publications Archives

that it would fit the more slender physique that resulted from his illness in Peru. And she began starching layers[12] of petticoats for herself and for Nelly—layers that had been impractical on shipboard but were now necessary accoutrements for the latest fashion. Then she packed personal family belongings, scrubbed the stateroom and the cabin, got out the muslin covering for Joseph's black tufted horsehair chair, and packed the multicolored pad for her black rocker. The settee, of course, was part of the ship, as were the storage drawers, and therefore little furniture would need to be shipped on to Nantucket. When July 4th brought a light wind from the west and beautiful weather, the crew commenced painting the ship. The third mate recorded, "Today is the anniversary of American Liberty—84 years ago—no time to celebrate." But most importantly, they were now north of Philadelphia—safe at last.

Finally, on July 12th, 1861, the *Aurora*, with all flags flying and glimmering under the bright sunshine of a summer day, entered its designated port of New Bedford. The first to board was neither the owner nor a city official, but a messenger boy[13] who ran up the gang-plank with a large parcel wrapped in brown paper—a new dress and bonnet for Malvina Marshall. At four o'clock, the Marshall family came ashore, looking quite splendid in their finery. Joseph, in dark suit, tie, and top hat, escorted his "better self" down the gangway. She, dressed in a "black and brown striped silk day dress with white lace cuffs and collar, and a brown straw open-brimmed bonnet, was cinched small at the waist and appeared with a voluminous skirt puffed out all the way to the hem." Nelly wore eyelet pantaloons under her white cotton print dress, but in the wind, "she could not keep her new straw hat on her head because the plaid taffeta ribbons came undone." This is how she later recalled her homecoming:

> *The first thing I remember after our stay in the hotel in New Bedford was for me the most distressing revelation—that Mr. Hussey would leave by the first train for his home in Fall River and that I should not see him again. My playmate and companion for five years—not to see him again! Why couldn't he come with us? Why must he go off away by himself? I was inconsolable. We were going home to see the aunts and uncles and cousins and he wanted to see his people. But he could go for a few days couldn't he and then come to wherever we were? [It was] No use. All the pleadings were useless and after a torrent of tears I fell asleep.*

Malvina Marshall's silk and cotton dress. *Worn at her New Bedford homecoming in July 1861, the brown and black striped dress has lace trim with diamond-patterned velvet at the sleeves.* – Nantucket Historical Association

Joseph and Malvina would be united with family again and spend the winter months at number 46 Fair Street, and from spring to fall, at Bloomingdale Farm. Nelly had little idea what her life ashore would be like, and her only tangible connections to her life at sea were her pets from Talcahuano—a goat and a canary. There would, of course, be school with the noted Quaker teacher Hepsibeth Hussey, snow and icicles that magically turned into water, rides in the hay wagon, learning to drive Major—the pretty sorrel that pulled the family buggy—making hearts and rounds with her Grandmother Pinkham, and playing with friends her own age. In November, she would be ten years old, with a whole new life ahead of her.

REPORT

—OF—

Attendance, Deportment, Scholarship,

—AND—

AVERAGE RANK IN CLASS.

Statistics for Term ending January 16th, 1869.

NAMES.	Average.	Scholarship.	Deportment.	Attendance.	Absences ex. do. unexcused
FIRST CLASS.					
Lottie E. Coffin,	10	10	10	10	0—0
Helen Marshall,	10	10	10	10	0—0
Charles G. Pratt,	9.99	9.97	10	10	0—0
Minnie Smith,	9.99	9.97	10	10	0—0
Mary S. Coffin,	9.98	9.99	9.94	10	0—0
Alfred M. Allen,	9.98	9.95	10	10	0—0
Eunice S. Barney,	9.97	9.95	9.94	10	0—0
John C. Easton,	9.70	9.31	9.94	9.82	2—1

NAMES.	Average.	Scholarship.	Deportment.	Attendance.	Absences ex. do. unexcused
SECOND CLASS.					
Lizzie F. Mitchell,	9.98	9.94	9.99	10	0—0
Susie E. Hussey,	9.97	9.93	9.99	10	0—0
Phebe A. Meader,	9.95	9.88	9.98	10	0—0
Lelia C. Murphey,	9.96	9.87	10	10	3—0
Anna C. Swain,	9.88	9.66	10	10	0—0
Clara L. Manter,	9.81	9.83	9.70	10	0—0
Kate S. Hallett,	9.76	9.94	9.31	10	0—0
Sarah C. Robinson,	9.75	9.83	9.41	10	0—0
Annie W. Bodfish,	9.72	9.89	9.26	10	0—0
Charles D. Coffin,	9.64	9.64	9.27	10	0—0
Frederick H. Coffin,	9.06	9.60	7.75	9.82	5—1
*Thos. S. Sayer, Jr.,	9.01	9.39	7.82	9.82	0—1

Scale of rank from 0 to 10.

*Left before end of Term.

Nantucket High School report, 1869. – Nantucket Historical Association

Epilogue

Helen "Nelly" Marshall (November 26, 1851–October 18, 1939) graduated at the head of her class from Nantucket High School and was admitted to Vassar, where she became a student and protégé of famed professor of astronomy Maria Mitchell. She made several extended tours of Europe, during which she studied French and Italian, and after a visit via rail to her uncle in San Francisco, she traveled across Panama from Pacific to Atlantic—this time by train. She taught at high schools in New Hampshire and Connecticut, and as a life-long student, continued her education with three years of special study at Harvard and summer graduate seminars at Bryn Mawr College. She retired after forty-two years as a member of the teaching faculty at Norwich Free Academy—thirteen of which were spent as curator of Slater Memorial Museum in Norwich, Connecticut. She made her home at 71 Williams Street in Norwich, Connecticut, and returned often to Sans Souci, her family retreat on Nantucket.

Malvina Fitz Alan Pinkham Marshall (November 1820–July 20, 1885) devoted herself to the care of her mother, was active with Joseph in the Unitarian Church, helped Elizabeth and William on the farm, and belonged to the Nantucket Literary Society. She often traveled to Boston and Providence to visit her sisters and attend lectures, concerts, and political debates. After retiring from the sea, Joseph Marshall (December 31, 1811–October 11, 1879) enjoyed owning and riding the best and fastest horses on the island and served as a fire ward and as a member of the Nantucket School Committee. He continued to be an indulgent father to Nelly by financing her many foreign travels.

George Hussey (1835–1901) was married shortly after his return on the *Aurora*, and the Marshall family, including Nelly, attended the wedding in Fall River. After becoming a ship's captain, he was an active member of the Masters and Pilots Association and the Mariner's Society. Captain Hussey later retired from the sea and moved to Providence. According to the *Inquirer and Mirror*, an arrangement of ferns and pinks held an honored place at his funeral service in 1901—a loving memorial from Helen Marshall.

George Edgar Mills (1833–1912), after serving for three years as mate on the *Aurora*, sailed on the *James Loper* to New Bedford and married his fiancée, Mary Frances Hopkins. A native of Grafton, Vermont, he served in the U.S. Navy during the Civil War, was twice wounded in the line of duty, and was Master of the *U.S. Grand Gulf* and Aide de Camp for General John Newton. In 2017, two of his drawings were magnificently reproduced, and his poem, "A Sperm Whaling Song," was published in *O'er the Wide and Tractless Sea* by Michael P. Dyer. In addition, Mills' many lyrics for sea chanteys were published by the Kendall Whaling Museum and are archived at the New Bedford Whaling Museum.

Notes

Abbreviations

I & M – The Nantucket newspaper, *Inquirer & Mirror* (1821 to present), first published as the *Inquirer.*

NHA – The Nantucket Historical Association, all owned artifacts.

NHARL – The Nantucket Historical Association Research Library, consisting of written materials such as records, letters, documents, and books.

NBWM – New Bedford Whaling Museum, custodian of written materials, paintings, photographs, artifacts, and exhibits.

TTHH – *Through the Hawse Hole: The True Story of a Nantucket Whaling Captain* by Florence Bennett Anderson, a major source of Pinkham and Marshall history and identification of artifacts.

MemoirHM – *My Childhood on a Whaler* by Helen Marshall, unpublished. Helen Marshall Hall Brown, private collection.

MemoriesMM – Journal entries and notes by Malvina Marshall of her two Pacific voyages. Private collection, Brown.

LSQ – *The Log of the Sea Queen*, Joseph Marshall, master, kept by Benjamin A. Freeman, Oct 1851–April 1855, New Bedford Whaling Museum.

JA – *The Journal of the Aurora* kept by George Edgar Mills, in the archives of the reading room, New Bedford Whaling Museum.

HMHB – Helen Marshall Hall Brown, private collection.

Chapter One

1. **Malvina Pinkham Marshall**, the second wife of Joseph Marshall, was thirty years old when she left New Bedford on the *Sea Queen*, pregnant with their first child together. Joseph already had a son, Joseph W. Marshall, by his first wife, who died when the boy was very young. While it was not common for women to go to sea with their husbands, by the mid-nineteenth century more women were choosing to do so and more children were born at sea or on islands in the Atlantic and Pacific.
2. **The *Sea Queen*** made nine whaling voyages during her time. In 1888, after leaking nine inches of water per hour, the bark put into Port Victoria in the Seychelles and was condemned. A share from the voyage's 1,780 barrels of oil and from the auctioned vessel and its contents helped pay expenses for the non-American crew members who were unprotected by the American Consul.
3. **The *Sea Queen's* send-off** from New Bedford was recorded in detail in a letter from Malvina's sister, Elizabeth Pinkham Crosby.
4. **The *Sea Queen's* quick reaction to the impending storm** was later told by Malvina to her family and was also recorded by the mate. Gimballed beds were common improvements on the whalers in later years, see Leavitt, 2013, p. 40. Malvina's sea sickness was no doubt extreme since she was pregnant and not used to being at sea during the first weeks, when even the sailors experienced bouts of nausea.
5. **The *Sea Queen* and the *Aurora*** were built at Matthews, Mashow & Company in Dartmouth, Massachusetts. Alonzo Matthews had trained as a carpenter and John Mashow, taught by Laban Thatcher, was a free black man born into slavery in North Carolina.
6. **The quotations and paraphrases attributed to William Snow** are by the ship's mate, 1851 LSQ. The quote by Joseph comes from his first letter home to William and Elizabeth Crosby. It includes his descriptions of the storm and the damage to other ships, including loss and damages at the Azores, MS283, folder 9 no. 1. This letter, from November 28, 1851, refers to a previous writing, which is missing. It would have announced the birth of Helen on November 26.
7. **A description of the *Sea Queen* leaving New Bedford**, send-off attendees, and the people who rode out on the pilot boat can be found in letters from Elizabeth Crosby to Malvina Marshall, MemoirHM, p. 1. Joseph's letter from Horta records the conditions of the state room during the storm, MemoriesMM.
8. **The Horta Bay section from Russell and Purrington's** panorama shows the whaleship *Charles W. Morgan*, just right of center, the last American whaling ship still afloat—now docked at Mystic Seaport in Connecticut.
9. **Russell and Purrington's panorama, "Whaling Voyage 'Round the World," 1845–48.** In 1841, at the age of 37, Benjamin Russell left his wife and three children to spend the next three-and-a-half years aboard the whaleship *Kutusoff* on a voyage around the world. He hoped to earn enough money to clear debts suffered in the banking crisis of 1832–33. He earned $894.51, barely enough to pay his debts, but the trip inspired him to begin a career as a marine artist. He transformed the sketches he made at sea into lovely watercolors and oil paintings. Later he teamed up with local sign painter Caleb Purrington, and together they produced the 1,500-foot panorama entitled "Whaling Voyage 'Round the World," an extravaganza, even in its day. Text from NBWM exhibit.

Chapter Two

1. **The entry into the Bay of Horta, Fayal** is recorded in the ship's log, LSQ, and explains the change of the mates in charge.
2. **A gam was a visit between two ships at sea.** A gamming chair was used to transport women from the deck to the boat below and across to a visiting ship. It was also used to deliver the captain's wife to a dock or pier. Gamming chairs were constructed in a variety of ways, including a cut-out from a barrel.
3. **A ship's flag was used to drape only the captain's wife** when she left the ship.
4. ***The Dabneys: A Bostonian Family in the Azores 1806–1871*** identifies one hotel, The Silva, where American travelers would have stayed. However, no name was stated in any of the Marshall documents. Sedan chairs, similar to the ones used in Portugal and other parts of Europe, were used in Horta.
5. **Cold, fresh drinking water** was a valued commodity.
6. **Medical guides were a common inclusion on board ship** and included instructions on childbirth. The copy Joseph Marshall used was published in Boston in 1848 and had his name written in pencil on the cover. It can no longer be found.
7. **The birth of Helen Marshall** must have been premature. The Marshalls expected to be in the Pacific before the baby's arrival, and the birth weight, although not mentioned, was very small. Anecdotes from other Nantucket families often mentioned "seven-month babies" who had catching up to do. The ritual of tying off the cord was done as much to save the sheets from blood as it was to prevent infection. Handing the

child off to the mate has been discussed in reference to other births at sea. While details of Helen's birth were described in Joseph's and Malvina's letters to family, almost no mention of the birth was included in the ship's log.

8. **Malvina's return** to the *Sea Queen* after giving birth is reported in the ship's log. All other documents report that Malvina stayed in Horta, as this book describes, until she sailed in February on the *Io* for Boston.
9. **Descriptions of the house in Horta on the Azorean island of Fayal** mostly come from notes Malvina wrote down so that she could describe the house, and the city itself, to her family when she returned to Nantucket. According to Luis Manuel Machado Meneses, Director of the Horta Museum in Fayal, the descriptions of the city's gardens and streets and houses appear to be accurate. In 1851, kitchens within houses would have been rare due to fire risk, as buildings often lacked running water. Outdoor kitchens can be seen in old drawings of Horta, suggesting they were common. It is thought that the hydrangeas on Nantucket come from the Azores.
10. **The Azorean basket** used as Helen's first crib is in the Nantucket Whaling Museum; so is the black Nantucket-made rocking chair. The glass courting candle (lamp) is in the private collection of Bethny Brown Avila.
11. **Explanation of quote in Joseph's letter to William and Elizabeth Crosby.** "She has been to the World's Fair after her instead of, (simply) Boston." He is saying that Malvina's delivery was as long and arduous as a trip to the World's Fair and Chrystal Palace in London, rather than a quick sail to Boston.
12. **Joseph Marshall's first wife was Phebe Whippey,** who died shortly after their son Joseph W. Marshall was born in 1841.
13. **Coarse white bread with a flour-dusted crust** is sold today in Nantucket as Portuguese bread.
14. **Captain Seth Pinkham** was a close acquaintance of Charles Dabney, son of John Bass Dabney, who was appointed by Thomas Jefferson as the first American consul to the Azores. This was told to Helen by Malvina many years later. Pinkham and Dabney met in Horta as well as Boston. This probably led to the social meeting between Malvina and Roxana Dabney.
15. **Paulo is described in Malvina's notes,** MemoriesMM. She also recorded the information about the cook's brother as the maker of the basket. This was in direct contrast to family "stories" that stated the basket was made by "native women," MemoirHM, p. 4. First, the people of the Azores were European and not indigenous to the islands, and second, basket-making was traditionally the work of men. "The ship that rocked the cradle" was a quote from Brown, M. H., p. 14; and MemoirHM, p. 4.
16. **Azorean baskets** are still made in the Azores and in Massachusetts by Manuel Azevedo of Taunton. The house interior description is from notes in MemoriesMM. The description of clothing and fashion in Horta in 1852 is from *The Dabneys: A Bostonian Family in the Azores 1806–1871*, and from 1870s photographs sent to the author from the Museum of Horta.
17. **Compare/contrast between Horta and Nantucket** was described in a letter from Malvina to Elizabeth as were the cook and the housekeeper. A letter from Malvina to her mother included the description of re-working her best silks and shopping for gifts to take home.
18. **The information about the visit with Roxana Dabney** comes from Malvina's list of notes in MemoriesMM.
19. **Hearts and rounds** were small white cakes cut from tin forms and were popular sweets at tea time in Nantucket—often served with a white sugar frosting. Malvina's H & R cutters are in HMHB collection.
20. **Paulo was hired to help Malvina on the *Io*** because girls in the Azores were kept at home and not available for child care or travel. MemoirHM, p. 4.

Chapter Three

1. **The trip on the *Io* to Boston**, Seth Jr.'s assistance, and the transfer to the Nantucket boat were recorded in the MemoirHM, p. 5. The role of women in Nantucket was different, Stanton, E., *History of Women's Suffrage*, 1985, p. 428; Brown, M., "Rhetorical Paradigms in American Women's Public Address," pp. 23–26. Abolition societies, anti-slavery societies, and equal rights groups and conferences are described in Booker, M., *Among the Stars—The Life of Maria Mitchell.*
2. **Lucretia Coffin Mott, distant cousin to Malvina** and to Maria Mitchell, was a daughter of a Nantucket whaling master. She became the first and foremost American woman leader and speaker on abolition and suffrage and was mentor to Elizabeth Cady Stanton and Susan B. Anthony. Booker, M., pp. 5, 12, 40, 65, 457. See also Booker, M., *Frederick Douglass*, p. 43; Reid, Ronald F., p. 369–374. For further understanding of this era, see Levi Coffin, Jr., *The Free Produce Movement and the Atlantic Silk Company.*
3. **Treloney (also Trelonia) Pompey's role in the Pinkham family** is described in TTHH, p. 201. Her own family is described in "Sampson Dyer Portrait of a Nantucket Mariner" by Elizabeth Oldham. See Historic Nantucket, Fall 2012, volume 63, no. 2., p. 19. Born Trelonia Dyer in the early 1800's in Nantucket, she was the daughter of Patience Allen and ship's steward and officer, Sampson Dyer. See Dyer portrait in NHA 2013.2.1.
4. ***Uncle Tom's Cabin* by Harriet Beecher Stowe** was published in 1852, and copies arrived in Nantucket the same week that Malvina arrived home in March 1852.

Chapter Four

1. **Statistics for men leaving Nantucket** as well as men leaving cities and states for the California gold fields are reported in numerous books and online journals including *Historical Text Archives, Crossing the Isthmus; The Isthmus Crossing, The Argonauts, Maritime Heritage,* and *Gold.* The conversion of whaling ships to cargo and passenger ships became common. Whaling masters also had to sail clear of ports that were linked with travel to San Francisco so that crew could not jump ship to get to the gold fields.
2. **The greatest number of letters during the voyage** of the *Sea Queen* was between Joseph and William and focused on business details, banking arrangements, and orders for food and dry goods. Elizabeth attached details of family news and sent packages of magazines and yard goods. Joseph sent directions for Malvina's travel to Peru. Most important, Malvina wrote a detailed narrative of her travels from Nantucket to Payta, Peru. The original is in NHRL 2015.05, dated July 21, 1852. A handwritten transcript made by Florence Farrier Hall in the 1950s is from HMHB.
3. **Handbills for the *Empire State*** steamship give detailed descriptions of the voyage, including menus. Other details of the steamship are from Malvina's letter to family at home, 1852.

4. **"Uncle John is Sick in Bed"** was a popular game for parents to play with young children and can be found in *Traditional Games for Children in England, Scotland and Ireland.*

5. **Malvina in her letter home** to thank all her family for their hospitality omits the name of her sister Mary Pinkham Plaskett and brother-in-law, Captain Henry Plaskett, because they were in route to China, Anderson, TTHH, p. 234.

6. **Anderson also describes 46 Fair Street**, which became 40 Fair Street a century later when Nantucket changed street numbers. Description of One Pleasant Street, Whipple, p. 9–10.

7. **Graham bread was thought to be essential** to healthy living and was the regimen of progressive nineteenth-century families who subscribed to the tenets of Grahamism.

8. **Joseph sent directions for Malvina's travel to Peru.**

9. **Concerning the trip across the Isthmus,** all personal descriptions and quotations are from Malvina's letter (1852) and MemoirHM, p. 5–8. The geography and description of the surroundings are from *National Geographic: Panama, Gamboa Rainforest*, p. 106, Chen and Yorente interviews, and HMHB's travel to Chagres River and Gamboa Rainforest Aerial Tram, Panama, 2016. The Las Cruses Trail was described in a letter cited in MemoirHM, p. 6.

Chapter Five

1. **Joseph's Payta** arrival described in his letter to Elizabeth and William, 1852.
2. **Malvina quote from** MemoriesMM, p. 8.
3. **Blackfish** is another term for a pilot whale.
4. **Holystone** is a piece of soft sandstone used for scouring decks of ships.
5. **Description of Kanakas** in MemoirHM, p. 10, 12.
6. **Description of life on the ship** in MemoriesMM, p. 11.
7. **Description of whaleboats:** Leavitt, ch. 11; Philbrick, ch. 3; MemoirHM, pp. 22–26.
8. **Description of processing oil:** Leavitt, ch. 11; Philbrick, ch. 3; MemoirHM, pp. 22–26.
9. **Joseph's diverse duties** are described in MemoriesMM; Joseph helped with the laundry: letter to Elizabeth, p. 3.
10. **Malvina's preparation for work,** described in TTHH, p. 199; war on cockroaches in MemoirHM, p. 5.
11. **Malvina on navigation:** Booker, p. 22, and MemoirHM, p. 26.
12. **"Greasy luck":** When the decks were greasy with whale remnants, the sailors saw this as a sure sign of future financial success. In the heyday of New Bedford's whaling industry (1830–1860), family and friends would gather at the docks to wish departing whalers "greasy luck." The expression bid the whalers a bountiful voyage, one that would grease the decks and crew with cakes of blubber and casks of oil. In time, the term evolved to mean good luck in general or good fortune.
13. **Forty-eight pounds of butter** for the sailor's larder is recorded in JA, April 16, no page recorded.
14. **Turpin, terrapin, or land turtles** quote from MemoirHM, p. 34.
15. **Report on mail exchange:** LSQ and envelopes addressed to "the Pacific Ocean."
16. **A "lay" is a percentage of the ship's profits.**
17. **Nelly losing her sea legs** refers to Nellie's adjustment to walking on land after being at sea. MemoirHM.
18. **Joseph's letter to William** mentions how he would like to see Main Street and [Nantucket] pine trees again, December 1854.

Chapter Six

1. **Joseph Marshall's sea chest,** photographed in 2018, is not gray because it was refinished in 1957. Malvina's oval top trunk is in the private collection of Bethny Brown Avila.
2. **Description of 46 Fair Street** and Mary Brown Pinkham's home interiors is in TTHH, pp. 173–176. The gold-banded white tea set is in the private collection of Bethny Hall Maury.
3. **Pinkham family information:** TTHH.
4. **Research on the Marshall family on Nantucket** and of One Pleasant Street was performed by Betsy Tyler using public records and and her published article (see biblio.). The mahogany bedroom suite is in the private collection of Elizabeth Marshall Steinbaugh. The Winthrop silver and coffee service is in the private collection of HMHB. For more on One Pleasant Street, see Whipple, pp. 9–10; and TTHH, pp. 212–214.
5. **San Souci** quotes and descriptions are from TTHH p. 176–177; in Betsy Tyler's *Sconset*, and in Patricia Butler, pp. 12–14.
6. **Descriptions of the Unitarian Church** are from Elizabeth Oldham, *Tale of the Bell*; Whipple, pp. 140–142; TTHH, pp. 204–207; Stevens, pp. 176–178; and Treloney Pompey TTHH, p. 201.
7. **Poem ending Chapter Six is from** TTHH.

Chapter Seven

1. **Descriptions of the *Aurora*:** The name "Aurora" is from Greek mythology—see *Myths and Legends of the World: Aurora.*
2. **The new design** of the crow's nest is from Joseph's letter to William, 1856.
3. **Nelly's pea jacket** is described in MemoirHM, p. 27.
4. **Decks cleared for laundry** is from JA, p. 12.
5. **The Dec. 13 gam** with Captain Jeremiah Ludlow of the *William Hamilton* is recorded in the JA, p. 12.
6. **Because these descriptions** on the *Aurora* come from a journal rather than a ship's log, there is more detail as to content and subject matter.
7. **At Recife, Brazil,** commodities and services did not come cheap. On his 1841 voyage aboard the New London whaleship *North America*, Francis A. Olmsted wrote, "We were obliged to pay six dollars for the hire of a boat [to ferry the men ashore], which is a good illustration of the exorbitant charges for everything we purchased at Pernambuco. A ship could be provided with supplies at the Sandwich Islands at a far cheaper rate than at this city, although they are six times as far removed from communication with the great commercial nations."
8. **The captain's delivering "a cantriver in respect to their lives"** is from JA, p. 21, and defined for the author by Michael Dyer of NBWM. Frequent notes in JA explain the navigational training provided for the officers by Captain Marshall, whose diligence prepared officers to become future captains and take command if needed.

9. **Dolls were made by Izannah Frankford Walker**, 1817–1888, in Central Falls, RI. Walker began a cottage industry of making dolls and doll furniture, employing several of her sisters to paint faces and make clothes.

10. **Helen makes note of cutting oranges** to aid in learning arithmetic in MemoirHM, p. 16. She learned to recite Longfellow's *The Village Blacksmith*, 1840, as part of her school work. Detailed notes from MemoriesMM and JA described first aid and health care on the voyage. Literature of the period included many health regimens such as exercise.

11. **The report of the storm conditions on the trip around Cape Horn,** including the direct quotation from George Mills, is from JA, p. 18. Often, grog was poured as a reward, in celebration of successful whale kills, or after safe delivery from severe storms. Grog was usually watered-down rum, sometimes with lemon juice added. Salted meat, sometimes horsemeat, was common fare on whaleships and resulted in health problems as reported by Mills.

Chapter Eight

1. **The singing and dancing on the *Aurora*** were in celebration of safe travels and described by Malvina in MemoriesMM and in MemoirHM, p. 33. A sailor's hornpipe was an athletic dance that evolved from Scottish, Irish, and English folk dances. It probably did not become a sailor's tradition until the mid-1700s. A century later, American sailors added new steps and "gymnastics" to the fast-paced dance, including the requirement to spit over the shoulder.

2. **John Oatley was the youngest crewman** and came from Smithfield, RI, JA, crew list.

3. **Shipkeepers were often** the last men to be called upon to climb aboard the whale boats or process the whale, Philbrick, pp. 34, 69. They were on par with low-ranking officers. It was necessary for all men aboard to have at least fundamental knowledge of sailing a whaler.

4. **Information on Valparaiso, Chile** is from *Historic Quarter of the Seaport City of Valparaiso*. Port facilities in Valparaiso included wharfs.

5. **Gam with the *Courier* of Nantucket—a near tragedy,** JA, p. 200; JA, p. 221; and MemoirHM, pp. 31–32.

6. **Malvina's tatting shuttle**, NHA collection 2013.2.3, MemoirHM, p. 32. Nelly's tattoo, MemoirHM, pp. 32–33. Marketplace quote, MemoirHM, pp. 19–20.

7. **Description of the Somontez house and family**, MemoirHM, pp. 19–20. Description of Señor Somontez' apparel, MemoriesMM, Druett 2001, p. 152.

8. **Demi-johns were large glass containers** used on ships for water storage. Often clear green and about 24 inches tall. Some were covered with rattan.

9. **Post office box on Charles Island in the Galapagos** used by whalers and other ships. Located on what is now named Floreana Island, the box is in use as of 2018 and requires no postage (see *National Geographic:* "World's Most Unusual Post Office," Sept. 18, 2017). Mail by ship was addressed with little detail and was remarkably reliable in delivery. "Joseph W. Marshall, Nantucket" was delivered from the Pacific mail box.

10. **The Marshalls' rented house in Lahaina, Hawaii** is described in MemoriesMM.

11. **The story of Nelly learning to swim** is from MemoriesMM.

12. **The description and story of Nelly meeting** the King and Prince of the Marquesas is from MemoirHM, p. 30.

13. **Entering Tahiti,** JA; MemoirHM, p. 29.

Chapter Nine

1. **To improve the speed of whale boats, the bottoms** were painted with a mixture of yellow ocher and paint oil, JA, Aug 24, p. 154.

2. **On Henderson Island,** Berry and Mills went ashore with trade goods, JA, Aug 23, p. 161.

3. **In Tumbez, Nelly received presents from home:** a new slate, a primping mirror (now in the private collection of Judith Hall Walker), and a miniature rope bed with mattress, NHA collection 1999.37.2a-b.

4. **Captain Marshall bought** whale guns for each boat, JA, Aug 23, p. 161.

5. **"[A]bout 35 barrels, 100 barrels, gams and six piglets born,"** MemoirsHM, p. 25.

6. **Mr. Hussey's desk,** MemoirsHM, p. 25.

7. **M.F. Maury, called the father of oceanography,** based his map on information—such as water temperatures, weather, and whale sightings and killings—provided directly from mariners and their logbooks. One whaler called it "a precious jewel." Maury himself regarded it as invaluable to whalemen whose assets "fishes up annually from the depths of the ocean, in real value of which far exceeds that of the gold mines in California." Maury centered the map on the Pacific, showing it as the primary habitat of whales, particularly the sperm whale.

8. **The first American ship,** the *Washington* of Nantucket, under Captain George Bunker, put into Callao in 1792. Since then, it had grown to be an important place where whaleships could be provisioned and repaired, barrels of oil sent home, and mail picked up and delivered.

9. **Nelly's near accident,** JA, p. 71.

Chapter Ten

1. **Malvina's night trip to Somontez Plantation.**

2. **Earthquakes** in South America were often accompanied by sea quakes.

3. **George Edgar Mills discussed his illnesses** in JA; notes of his illness by Malvina in MemoriesMM.

4. **Daniel Nevers of Martha's Vineyard**, 40 years old, was second mate on the *Aurora* and an experienced seaman.

5. **Regarding leak on the *Aurora* during trip to Callao, Peru**: bilge pumps and amount of ballast determined safe, according to statement by U. S. Navy Damage Control Training Center in Philadelphia.

6. **Ship chandlery and shipping agents of Crosby & Company,** Callao, Peru. See Crosby, Everett, p. 92, and MemoriesMM.

7. **Tea towels with red initial cross-stitch** made by Helen Marshall, NHA.

8. **A boat for Nelly,** see JA, p. 71.

9. **Historical information about the Civil War** is from Bailyn, Bernard.

10. **Tortoise shells as candles** on the water, MemoirHM.

11. **Joseph's personal thoughts about family** are in his letters home.

12. **Starching petticoats**: Malvina also added sugar-water to the starch.

13. **Delivery boy with the dress,** MemoirHM.

Acknowledgments

Dr. Larry Dodd, Director of Penrose Library at Whitman College deserves the highest accolades for his advocacy in transferring special collections from the Florence Bennet Anderson estate to the Nantucket Historical Association. I met with Dr. Dodd in October 1987 in Walla Walla, WA to discuss the Captain Seth and Mary Pinkham portraits, the Captain Joseph William Plaskett portrait, and the Pinkham desk. Returning two years later, I watched as the three portraits were wrapped and crated by the art department staff for shipment via 'Van Go" to Nantucket where they were received by NHA Administrator, John Welsh who organized a celebration of the Whitman gifts. Indeed, these two men impressed upon me the value of noteworthy collections and served as my inspiration to work toward additional acquisitions. I subsequently returned to Whitman in search of the desk, found it in a basement storeroom in the former Anderson home (now the Baker Faculty Center), and with Dr. Dodd's help, arranged for it to be returned to Nantucket. This time it was NHA Curator, Michael Jehle who received the desk, researched its origins, and planned for its exhibition in Nantucket and at Winterthur Museum in Delaware. The extraordinary generosity of time and guidance that Larry Dodd, John Welsh, and Michael Jehle gave to me must not be forgotten.

Yet, *primus inter pares*—first among equals—is Betsy Tyler, who over many years has been my advisor in all things research. This book was planned to be a brief children's story until I found that the Marshall family members were all of historical interest. Thus, it was Betsy Tyler's meticulous enquiry that made this present effort possible. Aimee Newell, Ph.D., formerly of NHA, was the first to publish an article on Malvina Marshall's dress and include the story of its delivery to the *Aurora (see bibliography).* NHA's Ben Simons always made time for my questions including an afternoon at the Folly at Squam Head where we discussed future acquisitions. Tony Dimitri took photographs for me and received donations with great care and professionalism.

Michael R. Harrison, NHA Curator, deserves a mighty thank you for his patience and continued support in helping me find artifacts, letters, and published information; and for his magnificent design and implementation of the 2017 exhibit, *Dear Ones: The Seafaring Pinkhams* in which the Marshall family story is featured. Also, Michael Harrison, the author of *Collecting Nantucket*, has presented Malvina Marshall's dress and Helen Marshall's ambrotype and jump rope.

In stating my appreciation, however, I must cast a wider net. The New Bedford Whaling Museum has become a new "home" due to the help and courtesy of Mark Procknik, head librarian in the museum's reading room. His ability in authenticating ships and dates and logs and journals has added to my book's credibility and worth. And when my energy waned, I needed only to talk to Sarah Budlong for a boost of new enthusiasm—especially if I mention the *Grand Panorama*. New colleagues across the water who have expanded my knowledge are Luis Manuel Machado Meneses,

Author Marsha Brown with husband Bob at the top of the Gamboa Rainforest tramway, overlooking the Chagres River. – Photo by Ikor Yorente

Curator at the Museum in Horta, Azores; Paulo Brasil city guide in Horta, and Marco Fernandes, my American tour guide in the Azores. I am grateful for the week I spent with my guide, Gustavo Schultz Chen in Panama, and my thanks to Ikor Yorente, my guide in the Gamboa Rainforest Reserve who pointed out the very place on the Chagres River where Malvina rented her mule for the Isthmus land trek.

I am indebted to those at Old Mystic Seaport who answered my questions and to the conserved exhibits—especially the *Charles W. Morgan*—that provided me with a closer reality to the age of whaling. David Iler demonstrated the use of quadrant and sextant, and Randi Green offered a detailed lecture on the contents and use of a whale boat.

My first manuscript readers—Robert Leach from Rhode Island, Nils Van Vorst from Massachusetts, and Doris Hildebrandt and Karen Dowsett from Toronto—were generous with their time and helpful with their comments.

Illustrator Dawn Secord survived my never-ending requests for corrections and details and succeeded in creating the perfect cover art for this book. Elizabeth Steinbaugh carried the Marshall chest to her Aurora Gallery in Vancouver, Washington to be photographed by Kate Singh. Tracey Papke and Celia Aldana, from Potamus Press, Murrieta, have been a vital source of talent in putting my design ideas into publish-ready form.

A sincere thank you to all my friends and colleagues at the Murrieta Rotary Club who continued to encourage my work on this book, and to Jann Gentry for photos of small artifacts. Kristin Sackett: I could not have done all this without you. Your diligence with a style manual and your patience with receiving copy that needed indents and formatting requires special thanks.

Peter Frost from Melbourne, Australia—chief pal in all my projects—shared my travels of discovery from the earliest Pinkham exhibits at the Peter Foulger Museum in Nantucket to the search of artifacts at Whitman College including afternoon tea at the college president's home where we found the long-missing Plaskett portraits.

With sincere thanks to the Spinner staff—Joe Thomas, Jay Avila, Susan Grace, Zachary White, and Katie Battaglino—for their expertise and personal attention to this project. And to Claire Nemes for the final editing of the manuscript.

Finally, thank you to my own "Dear One," Bob Brown, who has been patient with my long absences working on the computer, my requests for photographs, and the many places we needed to travel for research and better understanding of the Marshall family experiences. It was particularly exciting to have him by my side—from California to Nantucket, New Bedford, Plymouth, Providence, and Mystic; in the rainforests of Panama and at the wide blue bay at Horta, Azores.

– Marsha Hall Brown

Afterword

The *Sea Queen* of Westport, Massachusetts—the ship on which Malvina and Joseph took their first voyage together—was built in Dartmouth in 1851 by Matthews and Mashow Company for Andrew Hicks. It measured 100-feet-3-inches in length and 24-feet-one-half-inch in width. In 1852, while in the Pacific, Captain Joseph Marshall ordered a reconfiguration of the captain's cabin and the stateroom to allow for family living space. This included an alcove with a crib for baby Helen.

The *Aurora* of Westport—the ship that took them on their second voyage together—was also built in Dartmouth by Matthews and Mashow for Andrew Hicks, with Joseph Marshall as part owner. Bark *Aurora's* maiden voyage was launched from New Bedford on November 10, 1856. The ship was 112 feet in length with a 26-feet-4-inch breadth. To reduce confusion, there were three nineteenth-century ships named "*Aurora*." The others were a British packet and a whaleship that sailed the Alaskan-Pacific grounds.

In regard to the dates of events I have included in this account: If any of the dates I found in the ship's log or journal conflicted with those recorded in family letters, I have chosen to use the dates found in the contemporaneous letters. Where I have included content found in undated letters, I have approximated the dates based on careful research. Additionally, the exact sequence of the ship's arrival at various ports of call may not be precisely accurate, but is based on known facts about similar Pacific voyages. I was particularly surprised to find errors on the ship's crew lists and omissions of significant events that were recorded elsewhere.

You may have noticed that I chose to use the original spellings when compiling this story; these are the spellings found in letters or in Helen Marshall's memoir. The place names "Fayal" (Faial) and "Payta" (Paita), for example, appear herewith in the English form used at the time. "Chagres" has been corrected from "Shagres"—the way it appears in the memoir. I also use the journals' spelling of "Somontez," which appears elsewhere in forms such as "Somontes" and "Samontes." In addition, some punctuation has been modified for ease of reading.

I have, in some cases, used oral reports (or stories) passed down through generations by family members as part of my research, but only after validating their accuracy. Other oral histories were deemed to be myths or simply could not be verified. One example is particularly interesting: family stories maintained that Helen Locke (Farrier) crawled out over the metal roof of One Pleasant Street so that she could meet her friends in town. In fact, children in the family were repeatedly taken to the house and were shown the roof where this "drama" had occurred. However, due to the detailed research on the house by Betsy Tyler, it was found that the dates did not match. Helen Locke never lived in the house, and it was Harriet Pinkham (Locke), Helen's mother, who was living with her sister and brother-in-law on Pleasant Street when the escape took place. This was later substantiated by a note to Malvina stating that "Mother was never told about Harriet on the roof."

A brief back-story about how this book came to be: Around 1915, Helen Marshall began writing *My Childhood on a Whaler*, a memoir of her younger days. Although some of it was based on what she remembered, she also used the large collection of letters, documented memories, and lists of events written by her mother and father (Malvina and Joseph), her Aunt Elizabeth, and other family members. As she was probably in her late 60s when she began the project, she likely would have been relieved to have her much younger cousin, Ednah Crosby Farrier, join the effort.

Ednah was working in the publishing field in New York City and was therefore familiar with typing and organizing a manuscript. But Helen Marshall died before any real publishing plans had been made. There is evidence that Ednah began editing the thirty-five-page manuscript, the twenty-page "Memories," and Malvina's collection of notes in subsequent years. Clear notations, page-change instructions, and paragraph corrections (presumably Ednah's), appear on the pages of the typed edition of the memoir. However, none of the letters were transcribed and no new research was noted; Ednah's project had come to an end before it could be published.

A third writer, Thomas W. Hall (the husband of Ednah's sister, Florence), also tried and failed to bring Helen's story to print. While a Ph.D. candidate at New York University in the 1920s, Hall had submitted to the chair of his thesis committee a three-page proposal for research and a plan for publishing the memoir. That effort merely added paperwork to a growing collection of artifacts for me to sort out later.

I became part of the Marshalls' continuing family history when I was born the year before Helen Marshall died; I am her namesake. Then in 1957, my mother, Florence Farrier Hall, gave me Joseph Marshall's sea chest and several years later, I was given Malvina's letter recounting her trip across the Isthmus of Panama with her baby daughter, Helen. Over the years, I have received other Marshall artifacts—many of which I have donated to the Nantucket Historical Association, where all members of the family and extended family may view them. It was not until my retirement from teaching that I began sorting through the contents of the many boxes of Marshall memorabilia with the purpose of writing a book. Although this will be my fifth publication, it has been the most difficult and time-consuming. Very few of the family's papers were organized, and some of what Malvina left were merely lists of things she planned to tell her family when she got home from her voyages. Also, only four of the Marshall letters had been transcribed; information in the ship's journals and logs was different from that which was recorded in letters and other documents; and family stories, when compared with known facts and dates, were sometimes incorrect or exaggerated.

Still, this has been a fascinating tour of discovery, and I am deeply indebted to my mother for her inspiration and for saving almost EVERYTHING, and to my brother, Tom for helping me tell the story. –MHB

Index

Lahaina, Maui, 1845. *Whaler at anchor in Lahaina Habor, Maui.* –From Russell and Purrington's panorama, "Whaling Voyage 'Round the World," New Bedford Whaling Museum

The Fiji Islands, 1840s. *Friendly natives venture out in fancy canoes and catamarans—their huts and temples nestled between a warm magenta sky and white sandy beach. The steamship at far right, sporting an eagle below its smokestacks, may be a vessel of the U.S. Mail Steamship Company or the Pacific Mail Steamship Company, both of whom competed to serve Fiji in the late 1840s.* –From Russell and Purrington's panorama, "Whaling Voyage 'Round the World," New Bedford Whaling Museum

A scene from Huahine, Society Islands, ca. 1845. *This excerpt from Russell and Purrington's panorama shows village huts and missionary settlements, a raft of casks being towed to a vessel, people swimming, and a whaleship nestled in a shady grove of palm trees.* – New Bedford Whaling Museum

Illustrations from the journal kept by George E. Mills aboard the *Aurora*. *Mills wrote: "Whaling Illustrated.... The bow boat [top] and its crew close to an eighty barrel sperm whale, Jan. 25, 1856."* – New Bedford Whaling Museum

Selected Bibliography & Credits

Books & Print Sources

Anderson, Florence B. *Through the Hawse-Hole.* New York: The Macmillan Company, 1932.

Anderson, J. *Outspoken Women.* Iowa: Kendall Hunt, 1984.

Ashe, T. *History of the Azores, OR., Western Islands.* (Vol. 1), 2012.

Bailyn, B. *The Great Republic.* Massachusetts: D.C. Heath and Co., 1985.

Baird, Donald. *Women at the Sea in the Age of Sail.* Toronto, Canada: Nimbus Publishing, 2001.

Balano, James. *The Log of the Skipper's Wife.* Maine: Down East Books, 1979.

Bartholomew. *Illustrated Atlas of the World: Third Revised Edition.* Reader's Digest: The Reader's Digest Association, 1987.

Booker, Margaret Moore. *Among the Stars: The Life of Maria Mitchell.* Nantucket, MA: Mill Hill Press, 2007.

Browning, Robert M. Jr. *From Cape Charles to Cape Fear, The North Atlantic Blockading Squadron During the Civil War.* Tuscaloosa: University of Alabama Press, 1993.

Bryant, Jennifer. *Lucretia Mott: A Guiding Light.* Michigan: William B. Eerdmans Publishing, 1996.

Church, Albert C. *Whale Ships and Whaling.* New York: Bonanza Books, 1938.

Cordingly, David. *Seafaring Women.* New York: Random House, 2001.

Crosby, Everett. *The Making of Nantucket: Family Lives and Fortunes in the Nineteenth Century.* New Bedford: Spinner Publications, 2018.

Darden, Genevieve. *Whaling Letters.* Descendants of Whaling Masters, 2003.

Darwin, Charles. *A Naturalists Voyage Round the World.* Engravings by Evremond de Berard from the Atlas of Venus, Galapagos notebook, Vol. II, Issue 21, October 6, 1864.

D'Entremont, Jeremy. *Edward Rowe.* Massachusetts: Commonwealth Editions, 2004.

Draper, Charla L. *Cooking on Nineteenth-Century Whaling Ships.* Mankato, MN: Capstone Press, 2001.

Druett, Joan. *Hen Frigates: Wives of Merchant Captains Under Sail.* Simon & Schuster, 1998.

Druett, Joan. *Petticoat Whalers: Whaling Wives at Sea 1820–1920* University Press of New England, 2001.

Duprey, Kenneth. *Old Houses on Nantucket.* New York: Architectural Book Publishing, 1959.

Dyer, Michael. *O'er the Wide and Tractless Sea: Original Art of the Yankee Whale Hunting.* Massachusetts, Old Dartmouth Historical Society, 2017.

Early, Eleanor. *An Island Patchwork.* Boston: The Riverside Press, 1941.

Gallant-Simpson, Cynthia. *Sea Captain's Daughter.* South Orleans, MA: Cape Cod Cozies & Mysteries, Ink, 2009.

Harris, B. *Beyond Her Sphere.* Connecticut: Greenwood Press, 1978.

Harrison, Michael R. *Collecting Nantucket—Artifacts from an Island Community.* Nantucket Historical Association, 2018.

Jehle, Michael. *From Brant Point to the Boca Tigres: Nantucket and the China Trade.* Nantucket Historical Association, 1994.

Johnson, Andrew K., Roger Connor, Carlene Stephens, and Paul Ceruzzi. *Time and Navigation: The Untold Story of Getting from Here to There.* Washington D.C.: Smithsonian Books, 2015.

Leavitt, John F. *The Charles W. Morgan.* 2nd ed. Connecticut: Mystic Seaport Museum, Inc, 2013.

Littlefield, L.A. "Traditions of Padanarm" in *Sketches of Old Dartmouth, No. 2.* New Bedrod: Old Dartmouth Historical Society, 1903.

Lundy-Levy, Paula. *Tuck't In: A Walking Tour of Historic Prospect Hill Cemetery.* Nantucket, Massachusetts, 2012.

"View of the City of Talcahuano and the Port of Concepción," 1826. *In the 19th century, Talcahuano was well known to American whaleships as a place to replenish provisions and for crews to enjoy various forms of entertainment.* – From a painted lithograph derived from a Louis Choris original drawing, John Carter Brown Library, Brown University

MacKay, Anne. *She Went A-Whaling: The Journal of Martha Smith Brewer Brown...Ship Lucy Anne, 1847–49.* Oysterponds Historical Society, 1993.

Macpherson, Arthur George Holdsworth. *Mail and Passenger Steamships of the Nineteenth Century.* Philadelphia: J.B. Lippincott Company, 1929.

Mott, L. *"Discourse on Women."* In P. Kennedy & G. O'Shields, *We Shall Be Heard* (pp. 15–22). Iowa: Kendall Hunt, 1983.

Newell, Aimee. Piecework. March/April 2003. Fashions at Sea, Fashions at Home (p. 54–57).

Norling, Lisa. *Captain Ahab Had a Wife: New England Women & The Whale Fishery 1720–1870.* Chapel Hill: University of North Carolina, 2007.

Oldham, Elizabeth., Stewart, Paul. *The Tale of the Bell in the Unitarian Universalist Meeting House-A Scrapbook.* Nantucket. South Church Preservation Fund, 2015.

Philbrick, Nathaniel. *In the Heart of the Sea.* New York: Penguin Books, 2000.

Reid, Ronald. *Three Centuries of American Rhetorical Discourse: An Anthology and a Review.* U of Massachusetts, 1998.

Smith, S. "Address, Convention of Anti-Slavery Women," 1938. In K.K. Campbell, *Man Cannot Speak for Her,* Vol. 2. New York: Praeger, 1989.

Starbuck, Mary. *My House and I: A Chronicle of Nantucket.* Boston: Houghton, 1929.

Stevens, William O. *Nantucket The Far-Away Island.* New York, NY: Dodd, Mead, & Company, Inc., 1936.

Tyler, Betsy, et al. *Sconset: A History.* Nantucket Historical Association, 2008.

Whaling Letters. New Bedford: Descendants of Whaling Masters, Inc., 2003.

Whipple, A. *Vintage Nantucket.* Connecticut: Globe Pequot Press, 1978.

White, D. *Ain't I a Woman? Female Slaves in the South.* New York: Norton, 1985.

Yoakam, D. "Woman's Introduction to The Platform." In A.N. Brigance (Ed.). *A History and Criticism of American Public Address,* (Vol. 1). New York: McGraw-Hill, 1943.

Web Sites

Kelly, Caitlin. *Women in Atheneum History.* Atheneum Reference Library. http://www.nantucketatheneum.org.

Commentary Magazine. "The Impact of 'Uncle Tom's Cabin.'" Peter Wehner, July 8, 2011. https://www.commentarymagazine.com.

Essential Civil War Curriculum. "The Union Blockade of the Southern States, by Robert M. Browning, Jr." http://essentialcivilwarcurriculum.com/the-union-blockade-of-the-southern-states.html.

Leferve, Sherry. *Finders Keepers.* Nantucket Chronicle. 25 Jul. 2013. http://www.nantucketchronicle.com/wise-old-house/2013/finders-keepers.

Mabry, Donald. *Crossing the Isthmus of Panama before the Canal.* http://www.historicaltextarchive.com.

Merchant Marine in the Civil War. http://www.usmm.org/civilwar.html.

New Bedford Whaling Museum. "A Spectacle in Motion: The Grand Panorama of a Whaling Voyage 'Round the World." https://www.whalingmuseum.org.

Our History. Nantucket Unitarian Universalists. http://www.unitarian-churchnantucket.org/about-us/our-history.

Santos, Robert. *More History of the Azores: Early Accounts of Land Beyond.* Azores Web. http://www.azoresweb.com/history_azores_2.html.

Captain's daughter in a gamming chair. *A still-image from the film, "Down to the Sea in Ships," shows actor Marguerite Courtot (as Quaker Patience Morgan) being transported in a gamming chair from the bark* Wanderer *into a whaleboat. This is how Malvina was lowered from the* Aurora *into a whaleboat before being rowed to another ship. The movie is considered one of the most accurate film representations of whaling culture ever made. Shot in New Bedford in 1922, it employed the whaling ships* Charles W. Morgan *and* Wanderer, *and featured the only authentic whale chase ever recorded on film.* – Spinner Publications Archives

The Maritime Heritage Project. https://www.maritimeheritage.org.

The Panama Railroad. *History of the Panama Railroad Part II*. http://www.panamarailroad.org/history1b.html.

Westmoreland, Barbara. *On the Trail of Nantucket's Women's History*. "Historical Nantucket" magazine. Search on: www.nha.org.

"Whaling and Whalers." Galapagos Conservatory. http://www.galapagos.org/about_galapagos/about-galapagos/history/human-discovery/whaling. From Oxford, Pete, and Graham Watkins. *Galapagos: Both Sides of the Coin*. Morgansville: Imagine Publishing, 2009.

White, W. M. *A Brief History of the Galapagos: Darwin and Evolution*. "Galapagos Geologyon on the Web." http://www.geo.cornell.edu/geology/GalapagosWWW/Darwin.html.

World Catalog. https://www.worldcat.org. Search: "Marshall, Joseph 1811–1879" and "Logbook of the bark Aurora, 1856."

Unpublished Manuscripts

Brown, Marsha Hall. "And the Sea Rocked Her Cradle." Story of Helen Marshall. Collection of Helen Marshall Hall Brown. California State University Hayward, 1985.

Hall, Florence Farrier. "Whale Oil in Her Veins: the Story of Helen Marshall." Oral presentations to historical societies and high schools in Rhode Island, 1961–1980. Private collection of Helen Marshall Hall Brown.

Marshall, Helen. "Childhood on a Whaler." Interview by Ednah Crosby Farrier, 1920. Private collection of Helen Marshall Hall Brown.

Marshall, Helen. "Memories, 1880–1910." Private collection of Helen Marshall Hall Brown.

Marshall, Malvina. "Memories." Concurrent notes on voyages, people and Nantucket, 1856–1862. Private collection of Helen Marshall Hall Brown.

Tyler, Betsy. "Family of Joseph Marshall (1811–1879): Census, Probate, Real Estate, Genealogical, and Biographical Data," 2017.

______. "One Pleasant Street: A House History." Nantucket Preservation Trust: forthcoming.

______. "Time-line of One Pleasant Street." Outline of owners and dates.

Research Interviews and Conversations

Panama. Gustavo Schultz Chen, guide, Panama and Panama Canal. Ikor Yorente, guide, Gamboa Rain Forest Reserve, 2015.

Fayal, Azores. *Museu da Horta:* Luis Meneses, Secretaria Reginal da Educacao, Ciencia e Culture, 2016. Paulo Brasil, guide, 2017. Evangeline Garcia, guide, 2017.

Whaling / General. *Mystic Seaport:* David Iler, navigation topics, use of quadrant and sextant. *New Bedford Whaling Museum:* Mark Procknik, Michael Dyer. *Nantucket Historical Association:* Betsy Tyler, Michael Harrison, Elizabeth Oldham.

Exhibits

New Bedford Whaling Museum: On exhibit in 2018 was *A Spectacle in Motion: The Grand Panorama of a Whaling Voyage 'Round the World.* This panorama, painted in 1848 by New Bedford artists Caleb Purrington and Benjamin Russell, authentically depicts a whaling voyage originating from the port of New Bedford in the mid-19th century. Painted on fabric so that it could be presented on moving reels.

Nantucket Whaling Museum: On exhibit in 2017–2018 was *"Dear Absent Ones": The Seafaring Pinkhams*, curated by Michael Harrison, Obed Macy Director Research and Collections. This exhibition looked at Nantucket seafaring through the eyes of the Pinkhams, a local family whose sons and daughters went to sea over multiple generations. Includes numerous artifacts from Malvina Pinkham Marshall, Captain Joseph Marshall, and Helen "Nelly" Marshall. See Nantucket Historical Association Exhibit Archives online.

Archives

Brown University, John Hay Library: Carleton D. Morse Whaling Collection.
Brown University John Carter Brown Library: Whaling log books, journals.
Providence Public Library, The Nicholson Whaling Collection.
Nantucket Historical Association: Research Library.
New Bedford Free Public Library: log books, newspapers, genealogy.
New Bedford Whaling Museum: log books, art, panorama.
Spinner Publications: photograph and negative collection.

Nantucket Harbor, October 15, 1842. *An excerpt of a diorama depicting the* Napolean *(far left) alongside Straight Wharf, and the* Tyleston, *the* James Loper, *and the* Joseph Starbuck, *alongside North Wharf. In the distance are the* President *and the* Planter. – Nantucket Historical Association

Letters

Marshall, Joseph, to William and Elizabeth Crosby. November and December 1851. Horta, Azores.

Crosby, Elizabeth Pinkham, to Malvina and Joseph. October 1851. New Bedford.

Marshall, Malvina Pinkham, to family at Nantucket. December 1851. Azores.

Marshall, Malvina, to family. About trip to Panama and Peru. July 21, 1852. NHARL 2015.05 MS. Florence Hall, 1958.

Marshall, Malvina, to family. About trip aboard British ship to Peru. July 6 and July 21, 1852.

Marshall, Joseph, to William and Elizabeth Crosby. About leaving Peru. July 1852.

Nantucket Historical Association: Research Library

Collection 283, Box 1 ~ Marshall-Pinkham Papers/Brown Collection, 1850–1894.

- Folder 1, Letters no. 1, 2, 5, 6, 7.
- Folder 6 Letters 1, 2, 4.
- Folder 8, Letters 2, 3, 4.
- Folder 9, Letters 1 (from Fayal after storm and birth of Helen), 2, 3, 4, 5, 6 , 7, 9, 10 (Joseph's last extant letter to William before leaving for home Dec. 1860. pp. 1–4).
- Folder 17, Letter 1.
- Folder 19, Letter 7.

Collection 302 ~ Pinkham Family Papers/Mason Collection, 1841–1932.

- Folder 2, Letters 1 and 2.

Note: In the examination of the letters between the Crosbys and the Marshalls, it appears that when the contents included business arrangements and/or instructions, Joseph signed such letters with his full name (HMHB).

Image Credits

All photographs and artwork in this book are credited to the source from which they were obtained. If possible, the artist is identified in the caption accompanying the image. In the references below, the letters "a," "b," "c," following a page number refer to the image's location as it appears on the page, in order, from left to right, top to bottom.

Bancroft Library, UC Berkeley: 28

Brown Family Collecction: v; 66a-c; 67a-c

Brown, Robert S., photograph: 48

Brown University Library: 70

Gentry, Jann, photographs: 57; 67a-c

John Carter Brown Library, Brown University: 100

Library of Congress: 14; 20; 26; 27; 59a; 59b; 63; 71; 74; 81

Nantucket Historical Association: ii; 6; 12; 17b; 22; 43; 45a; 45b; 48; 53; 54; 57; 62; 67d; 79a; 85; 86, 102, 103, endleaves

National Library of Australia: 72

New Bedford Free Public Library: 11

New Bedford Whaling Museum: 5; 8; 10; 17a; 19; 39; 56; 64; 68; 69; 79b, 96-99, endleaves

Secord, Dawn, illustrations: Front Cover; 15; 29; 32; 61; 73; 82; 83

Singh, Kate photograph: 42

Spinner Publications Archives: 9; 18; 24; 30; 35; 40; 84, 101

Wikimedia Commons: 51; 60; 76

Yorente, Ikor, photograph: 92

Town of Sherbourne, Nantucket, ca. 1812. *"What wonder, then, that these Nantucketers, born on a beach, should take to the sea for a livelihood! They first caught crabs and quahogs in the sand; grown bolder, they waded out with nets for mackerel; more experienced, they pushed off in boats and captured cod; and at last, launching a navy of great ships on the sea, explored this watery world;...and in all seasons and all oceans declared everlasting war with the mightiest animated mass that has survived the flood."* – Text from Herman Melville's, *Moby-Dick*; painting by Thomas Birch, 1854, after a lithograph by Joseph Sansom, Nantucket Historical Association

133. Bark Sea Queen of Westport Pa.Oc.
Joseph Marshall
Wm F. Snow Nantucket 32 D Br.
Edward B. Coffin " 24 F "
Richard G. Luce Tisbury 20 L Br.
Randall M. Cornell Newport 22 D B
Andrew J. Sandsbury
Edward Burdett Nantucket 19 F Br.
Henry Malloy Boston 27 L L
Willis Bailey Saybrook 22 B Bk.
Chas. J. Mitchell Shirley 23 " "
Jno. Watts Tiverton 15 L Br.
William Martin Lancaster 22 " "
Richard Borden Taunton 21 " L
Horace Peabody Franklin 22 " Br.
Jas. Hamilton
Geo. Patterson
Wm Algeo
Jas. Wilson
Eben C. Clark Jr.
Edward A. Boyd
Brother Seth
many letters, but
tomorrow, I thought
as late as possible.
Wednesday we
12¢ Due
Received May 29
Mrs. Joseph
Bark Aurora, of
care of the U.S.
but the hours rode speedily before me
and again I was behindhand. But wait
half a moment I wish to write a few
to Nelly — My dear darling Nellie
commenced writing I think that the most
portant part of this should be written to you
We have a very beautiful little kittie and it
as if she almost knows that I am writing to
for she has come & perched herself upon the
played with the ink & is now carefully
my pen as it moves rapidly along towards
nose. She is our only pet so we have
red "Chenell" (cochineal) cord around her
there she has put her paw on the word
smootched it. I wish you could see her
is so very cunning you could not help